NEW MANAGEMENT

OF

ENGINEERING

Patrick D.T. O'Connor

www.pat-oconnor.co.uk

Also by Patrick O'Connor

<u>Engineering:</u>

Practical Reliability Engineering (4th. Edition), John Wiley and Sons Ltd., Chichester (2002).
www.pat-oconnor.co.uk/practicalreliability.htm

Test Engineering, John Wiley and Sons Ltd., Chichester (2001).
www.pat-oconnor.co.uk/testengineering.htm

In My Humble Opinion, Lulu Inc. (2005) www.lulu.com

<u>Fiction:</u>

Walter Minion's Therapy, Trafford Publishing (2004).

Walter Minion's Secret Life, Lulu Inc. (2005)

www.walterminion.com

Copyright © 2004 by   Patrick D.T. O'Connor
62 Whitney Drive
Stevenage SG1 4BJ
UK
(pat@pat-oconnor.co.uk)

This book was originally published in 1994 by John Wiley, Chichester, England, under the title *The Practice of Engineering Management: a New Approach.* It is re-published, with changes and additions, with the kind permission of John Wiley and Sons Ltd.

Published 2005 through Lulu Inc. <u>www.lulu.com</u>

Dedicated to the memory of

Alison Eadie

A wise and brave commentator on commonsense management

And of course to the great

Peter F. Drucker

And especially to my wife

Ina,

for giving me the last word

CONTENTS

PREFACE

The first edition of my book* did not have the impact that I had hoped it would. Sales in the first year were encouraging but then tailed off, and it went out of print after a few years. I have given much thought to the causes for this failure. I remain convinced that the fundamental message, as described in the original Preface, is correct. My years of experience since, my discussions with other managers and my reading of other management literature have not resulted in my altering or deleting anything, apart from the addition of new material to strengthen, support or update the message. The main reasons for failure, it seems to me, are:

- The people who teach management prefer, mostly, to teach across a range of disciplines, and not to specialise on engineering.

- The people who teach management are mostly not engineers, so they do not appreciate the major differences between engineering and other industries.

- Some of the ideas that I present are not popular or fashionable in academia.

- Good engineering managers are busy people, and they don't feel the need to read books to tell them how to do their jobs.

- I am an unknown in the field. To be commercially successful, books must be written by celebrities or academics teaching and researching in the field, the accepted "gurus".

There is, of course, another possible reason: the book might be superfluous, or even bad. But I repeat: I remain convinced that the central theme of connecting Peter Drucker's teaching to the particular challenges of managing engineering makes my book necessary and unique. So do other themes. For example:

---

* *The Practice of Engineering Management*, J. Wiley 1994

- It is still the ONLY book on engineering management (as far as I can discover; please correct me if you know otherwise) that discusses the subject of testing. Every real engineer knows that testing is the most difficult and most expensive part of any development and of many production programmes, but the writers and teachers seem to be unaware of this. (It is not taught on nearly all undergraduate engineering curricula).
- Quality and reliability (my specialist areas, so I might be accused of bias) receive scant attention in other books, despite the overwhelming revelations by W.E. Deming and the experience of the companies that have taken up the challenge, or been left behind.
- Real engineers know that ISO9000 is crap, but this uncomfortable fact is ignored by the successful authors.

Publishers are in business to make money, so, despite my efforts to have a second edition published, Wiley's (my original and excellent publisher, who have kindly returned the copyright to me) nor any of the other publishers I have approached, have been prepared to take it on. So here at last is my second edition. It won't make me money or fame, but if it helps any readers to deal with the exciting and important task of managing engineering I will consider the effort worthwhile. Good luck!

Patrick O'Connor

November 2004

ACKNOWLEDGEMENTS

I acknowledge with gratitude the help and guidance I have received, over many years of engineering, from excellent managers who have demonstrated the truth of the message I try to convey. I thank also all who have worked with me, and who have shown by their responses and criticisms that management works both ways, and that we need the gift to see ourselves as others see us if we are to improve.

I am grateful to friends at work who have provided ideas, help, advice and criticism. To (the late) Norman Harris at British Aerospace I owe thanks for many stimulating discussions on variation and on scientific method and thought. Barry Brown of British Rail provided necessary criticism of my chapter on development of engineers.

I must also thank the authors whose books have brought the great benefits of modem management to the world. Dr Peter F. Drucker is of course at the head of this list. I hope that I have succeeded in some way to bring more engineers to understand and apply the ideas they have taught.

I thank my family, and particularly Ina, for once again bearing the burden of a writer in the home, and for their love and support.

Throughout this book I use masculine prepositions to mean people of either sex. I apologise to any female engineers and managers who might be offended, but I hope that they will accept that the alternative of using expressions such as "him or her" throughout is tedious. Engineering has traditionally been an occupation for men, but I happily acknowledge the growing contributions of women.

QUOTATIONS

There can be little doubt that leadership in productivity and wealth during the second half of the twentieth century will fall to the country whose managers understand and practice management in its fullest sense.

Peter F. Drucker
The Practice of Management, 1955

We are going to win and the industrial west is going to lose out: there's nothing much you can do about it, because the reasons for your failure are within yourselves. Your firms are built on the Taylor model; even worse, so are your heads. With your bosses doing the thinking while the workers wield the screwdrivers, you're convinced deep down that this is the right way to run a business.

For you, the essence of management is getting the ideas out of the heads of the bosses into the hands of labour. We are beyond the Taylor model: business, we know, is now so complex and difficult, the survival of firms so hazardous in an environment increasingly unpredictable, competitive and fraught with danger, that their continued existence depends on the day-to-day mobilisation of every ounce of intelligence.

For us, the core of management is precisely this art of mobilising and pulling together the intellectual resources of all employees in the service of the firm. Because we have measured better than you the scope of the new technological and economic challenges, we know that the intelligence of a handful of technocrats, however brilliant and smart they may be, is no longer enough to take them up with a real chance of success.

Only by drawing on the combined brainpower of all its employees can a firm face up to the turbulence and constraints of today's environment.

This is why our large companies give their employees three to four times more training than yours, this is why they foster within the firm such intensive exchange and communication; this is why they seek constantly everybody's suggestions and why they demand from the educational system increasing numbers of graduates as well as bright and well-educated generalists, because these people are the lifeblood of industry.

Your 'socially-minded bosses', often full of good intentions, believe their duty is to protect the people in their firms. We, on the other hand, are realists and consider it our duty to get our people to defend their firms which will pay them back a hundred-fold for their dedication. By doing this, we end up by being more 'social' than you.

Konosuke Matsushita

1

INTRODUCTION

Peter Drucker's landmark book *The Practice of Management* [1] was published in 1955. In this book are to be found all of the profound ideas that have shaped the way that the world's best-managed companies and other excellent organizations work. Drucker exposed the poverty of so-called "scientific" management, which held that managers were the people who knew how all levels of enterprises should be run, and who should therefore provide detailed instructions to the "workers", who were assumed not to have the knowledge and skills necessary for managing their own work. They then had to manage the workers to ensure that they performed as required. "Scientific" management was the term used by the American engineer F.W. Taylor to define the doctrines he proposed during the early years of 20th. Century industrialisation. This management approach called for detailed controls and disciplines, and it inspired the production line, the division of labour, and emphasis on specialisation. Drucker showed that there is no level at which management stops: every worker is a manager, and every manager a worker. Modern workers have knowledge and skills that can be applied to the management of their work. Freeing these talents generates improvements in motivation and productivity that can greatly exceed "planned" levels. It follows that work involving high levels of knowledge and skill are particularly suited to the management philosophy presented by Drucker.

Drucker taught that work should be performed by teams of people who share the same motivations. Management's role, at all levels, is to set objectives, organize, motivate, measure performance and to develop the people in the teams. Drucker initiated management concepts which today seem new and revolutionary to many engineers, such as "simultaneous engineering", involving all the skills of design, production, marketing, etc. in an integrated development team from the start of a project. The "quality circles" movement, in which production workers are encouraged to generate ideas for improvement, is entirely in keeping with Drucker's teaching.

Drucker's teaching on management is universal. The concepts apply to the management of any kind of enterprise or organization, since they all depend for their success on the performance of their people. Therefore the ways that the people are organized, motivated, and trained are the crucial factors that determine relative performance. Drucker wrote that the people are "the only resource that really differs between competing businesses. Each business can buy the best machines, and their performance will not vary significantly between users. However, the performance of people, particularly as managers, can be greatly enhanced by applying the new first principles of management. Drucker wrote that the most important attributes of a manager cannot be taught or learned, but are inherent characteristics that must be sought out and developed. He forecast that countries whose managers understood and practised the approaches he described would become the world's economic leaders. Japanese managers quickly adopted them across nearly all industries.

It is an unfortunate fact that these ideas have not revolutionised management practice in the West to the extent that they did in Japan. In the West the ideas received patchy recognition and application: many leading companies owe their success to the application of Drucker's teaching, but other companies and organizations fall far short of their potential because their managers do not appreciate or apply the principles.

Many managers would regard most of Drucker's teaching as commonsense, and not revelatory. To the extent that they apply the ideas, this is a reasonable attitude. It explains why many organizations are managed very well by people who might never have heard of Drucker, or have not read "The Practice of Management" (or any of Drucker's later books). It is arguable that most of them would become even better managers if they did study Drucker.

Other managers instinctively tend to favour the ideas enshrined in "scientific" management. It is not uncommon for organizations to contain managers with different attitudes, leading to disagreement on the fundamental issues of how people and their work should be managed and to confusing changes as the different philosophies vie for ascendancy.

Ironically, engineers often have difficulty in applying Drucker's principles, yet the principles of "scientific" management, that managers manage and workers do what they are told, is fundamentally inappropriate to even the simplest engineering tasks. In fact, because of the pace of technological change, it is common for engineering managers to be less knowledgeable than many of their subordinates in important aspects of modern product and process development, so making a philosophy based on trust and teamwork even more necessary.

The main reason why engineers have tended to gravitate towards the "scientific" approach to management is that they are normally, and have been taught to be, rational, numerate and logical. Engineering is the application of science to the design, manufacture and support of useful products, and scientific education is rational, numerate, and logical. Therefore the ideas of "scientific" management were welcomed by engineers, and they have difficulty in giving them up in favour of methods that seem vague, subjective, and not amenable to quantification and control. This attitude is reinforced by the fact that few engineers receive training in the new management principles, and in fact much management training and literature is tinged with Taylorism. However, all engineering work is based on knowledge, teamwork and the application of skills. Applying "scientific" plans and controls to such work takes Taylor's original concept far beyond its original intent of managing manual labour. "Scientific" management, and related forms of organization and project control so often observed in engineering are inappropriate, wasteful and destructive of morale, both within enterprises and in the societies in which such principles are applied.

This book therefore attempts to relate Drucker's teaching to engineering, in the hope that many more engineers will understand and apply the simple, beautiful and rewarding concepts that Drucker explained so clearly. For those who have not read *The Practice of Management* I recommend that they do so before reading the chapters that follow*.

Who are the Engineers and Managers?

For the purposes of this book, an engineer is any qualified person engaged in engineering. These might be chief executives of engineering companies, people who market engineering products or services, people working in test labs or production lines, scientists and mathematicians working on engineering projects, as well as engineers involved in design, development, maintenance or any related activity.

---

* Of course Drucker has not been alone in writing excellent books on management. Others are referred to later. However, none of these addresses the specific and unique needs of engineering.

All of these people must share two qualifications: they must understand the scientific and engineering principles of the work involved, and they must be able to manage their part of it, including the work of the people for whose performance they are responsible.

Such a broad definition means that nearly every professional or technical person working in engineering is both an engineer and a manager. The engineering role is concerned with the product and the processes. The management role is concerned with the people in the team. The two roles must never be separated; separation almost always leads to bad engineering and bad management.

An engineer usually knows his strengths and limitations, since they are based on measurable performance in examinations and at work. Not every engineer will claim to be excellent, and most will admit to their weaknesses in terms of technical knowledge or experience, and will acknowledge technical mistakes. However, few are as aware and as frank about their performance as managers. Just as it is normal for people to consider themselves to be "right" in other subjective areas such as personal relations and politics, so most people are convinced that in their role as managers they are doing the right things.

Individual management styles are naturally as diverse as the personalities involved. Yet there is only one "right" way to manage, to which all managers must strive. Unfortunately many engineers find this way difficult to reconcile with their rational, numerate backgrounds, and with much of the management training they might have received. They treat their people as they would their machines and their computers: they specify the work to be done, maintain them well and control them. This is analogous to a football coach telling the players before the game how to play every move, then calling changes from the sidelines. This approach would inhibit both individual and team initiative and cooperation. On the other hand, a team that is trained and motivated, then given general as opposed to specific directions, will be much better able to take advantage of tactical opportunities and to respond to developments as they occur on the field. A similar parallel could be drawn with military situations: battles can be lost when the strategist interferes with decisions of the commanders in the field.

(Interesting comparisons can be made between the game of football as understood by Americans and as understood by others. American football is to a large extent a series of set pieces, planned in detail, and organized from off the field. The other kind of football, or soccer, like ice hockey, basketball and several other team games, is a game of great tactical freedom, the epitome of combined individual and team effort.

Sporting analogies will be used in later chapters. The un-American way of football will be the model used. It is also interesting to speculate on the influence of American football on American management practices, and *vice versa*.)

Engineering is Different

Managing engineering is different to managing most other activities, due to the fact that engineering is based on science. It is a first principle of management that the managers must understand the processes they are managing. Most non-scientific endeavours, such as retailing, financial services and transport planning can be learned fairly quickly by people with basic knowledge and reasonable intelligence. However, engineering requires proficiency in relevant science and mathematics and their application, and this can be obtained only by years of specialist study and practice.

Every engineering job is different to every other, even within a design team or on (most) production lines. Every one requires skill and training and there is always scope for improvement in the way they are performed. Nearly all involve individual effort as well as teamwork. These aspects also apply in different degrees to many other jobs, but engineering is unique in the extent to which they are relevant and combined.

Engineering is also different due to the reality that engineering products must proceed through the phases of design, development testing, manufacture and support. This is also true in part for some other fields of endeavour: for example, a building or a civil engineering structure like a dam or a bridge must be designed and built. However, these projects do not share some of the greatest challenges of most engineering creations: the first design is usually correct, so there is little or no need to test it. They are seldom made in quantity, so design for production, managing production and item-to-item variation do not present problems. They are simple to support: they rarely fail in service and maintenance is simple.

Engineering in a Changing World

Engineering is a profession subject to continual and rapid change, due to developments in science and technology, components, materials, processes and customer demands. It is also subject to economic and market forces and often to pressures of competition, so costs and timing are crucial. In particular, there are few engineering products that do not face worldwide competition, whether they are produced for specialists or for the public. Engineering managers must take account of all of these factors, scientific, engineering, economic, markets and human, in an integrated and balanced way. They must also balance short term objectives with long term possibilities, so they must be able to weigh the advantages and risks of new technologies.

No other field of management operates over such a wide range or in the face of so much change and risk. It is not surprising that many organizations have made the mistake of separating the management of people from that of technology, instead of facing the challenge posed by what Drucker called "the new management".

The chapters that follow will show how the new management philosophy should be applied to the art and science of engineering. This includes management of engineers and other project resources, of methods for design, test, manufacture and support, and of the scientific, technological and business opportunities and risks. The management of other aspects important or unique to engineering, such as research, variation, quality and safety is also explained.

The philosophy and methods described will not guarantee success, particularly in competitive situations. As in sport, the best are more likely to win. As in sport, there is also an element of luck: a good project can fail due to external forces such as politics or global economic changes, or a simple idea might be fortuitously timed to coincide with a market trend or fashion. However, again as in sport, there is little chance of success if the approach to the business and its practice is not of the best in every way. There are no minor leagues or second divisions in engineering business, and no amateurs. Survival depends on being able to play at the top. The philosophy and methods described in the chapters that follow have been proven to provide the basis for winning.

2

FROM SCIENCE TO ENGINEERING

The art of engineering is the application of scientific principles to the creation of products and systems that are useful to mankind. Without the insights provided by scientific thinkers like Newton, Rutherford, Faraday, Maxwell and many others, engineering would be an entirely empirical art, based on trial and error and experience, and many of the products we take for granted today, such as aircraft, computers, and dishwashers, would not be conceivable. Knowledge and a deepening understanding of the underlying scientific principles, often as a result of scientists and engineers working as teams, drives further development and optimisation across the whole range of engineering. Automobile engines provide a good example, in which improved knowledge of aerodynamics, materials and combustion processes have been combined with improvements in processes such as casting, machining and surface treatment, to result in engines that are smaller, lighter, more efficient and powerful, more reliable and longer lasting, and with lower undesirable exhaust emissions. The enormous range of modern electronic products is the result of applying the sciences of crystallography, atomic diffusion, particle beam physics, chemical vapour deposition, ultrasonics and many others to the fabrication of low-cost, incredibly complex integrated circuits. The scientists could not design the circuits, but without the understanding provided by the scientist, the design and process engineers would not be able to start. By itself, the theory has no practical utility, until it is transformed into a product or system that fills a need.

Engineers provide the imagination, inventiveness, and other skills required to perceive the need and the opportunity, and to create the product.

In its early days, engineering was a relatively simple application of the science that was known at the time. There was little distinction between science and its engineering application. The steam engine followed quickly from the realization of the energy available, and the electric telegraph was a simple application of electrical conductivity. These early developments usually involved a single discipline, or simple combinations of disciplines. Designs were not refined or complex, and manufacturing processes were simple. The products were also easily understandable by most people: anyone of reasonable intelligence could see how a steam engine or electric telegraph worked, and they were described in children's encyclopedias. Today, however, most products of engineering effort involve multi-disciplinary effort, advanced technology related to materials, processes, and scientific application, and considerable refinement and complexity; Most people cannot understand the electronic control system of an electric power tool, the principles of a laser disc recording system, or the stress calculations for a turbine blade. This complexity and refinement have been driven by advances in science and in its application, and by the ceaseless human motive to improve on what has been achieved before, by ourselves, by our competitors or by our enemies (real or imagined).

Engineering is also based on mathematics. All designs and processes involve evaluation, optimization and measurement. It is not possible to develop and produce control systems, communications, radars or vehicle suspension systems without the use of basic mathematics as well as more advanced methods such as differential calculus, Fourier analysis and the mathematics of mechanics, communications and control theory.

Engineers must understand and apply the scientific and mathematical principles that form the foundations of their work. All engineers are taught these as part of their basic and specialized education. In most practical situations engineers do not need to work with scientists and mathematicians, as they have sufficient knowledge to deal with the problems concerned.

SCIENTIFIC THINKING

The role of science is to determine the causes of the effects we observe in nature. We observe the universe, the world the world, the materials that constitute the Earth, effects such as light and heat, and we observe that connections exist: we rub two sticks together and we obtain fire and heat. Science is mankind's curiosity in action to discover the truths that explain the observations and the connections. Pure science seeks only explanations: it does not make or change anything, and it accepts that the basic laws that it discovers are immutable, and are true everywhere.

New discoveries in science, particularly the breakthrough theories like those developed by pioneers such as Newton, Darwin, Maxwell, Planck and Einstein, are usually the results of inductive thinking. Induction is the thought process by which ideas are derived from existing knowledge, but where the new idea does not follow directly from the knowledge. In some cases the connection between the idea and existing knowledge might be very tenuous. For example, Newton observed gravity and basic mechanics in action, and postulated the laws of motion. These laws are not obvious direct consequences of the observations, and a rare power of inductive thinking was required to make the connection. Likewise, Galileo's realization that the Earth revolves around the Sun could in principle have been determined by anyone else who observed the sky, but in practice it required a stroke of genius to interpret the observations in a way that was different to that generally accepted at the time. (In Galileo's case courage was also essential, since it was a capital offence to state publicly that the Earth was not the centre of the Universe, and Galileo eventually paid the price). Einstein's revelation of relativity and Planck's elucidation of quantum physics were further examples of induction, in which the concepts represented great mental leaps from the known world.

Whilst the thinking that leads up to an induction might be very long, the actual inductive thought occurs instantaneously: it is the moment when Archimedes shouted "Eureka". The inductive thought need not even occur while thinking about the problem. Inductive thinking is a uniquely human quality. We do not understand how inductive thoughts occur, so we cannot forecast them or plan for them.

It is interesting to note that inductive revelations in science are nearly always much simpler to understand than the groping theories that occupy the minds and time of the great majority of scientists. Inductive insights nearly always involve simplifications of existing theories, as well as clarification of causes and effects that the previous theories could not explain.

Science also proceeds by deduction. Deductions are thoughts that flow logically from existing knowledge or theories. Much of scientific progress is made by deduction based upon the main inductively derived theories. For example, "genetic engineering", in which genetic material is deliberately altered to develop plant or animal types with controlled characteristics, is based upon deductive reasoning, using the knowledge of inductive revelations such as Mendeleev's discovery of the periodic table of the elements, Mendel's discovery of genetic function and Watson and Crick's discovery of the double helix form of the DNA molecule.

Another form of scientific thinking lies between pure induction and pure deduction. In hypothetico-deductive thinking a theory is postulated, then tested against the evidence of knowledge. Modem particle physics seems to be dominated by this approach. No one is quite sure what lies inside particles such as electrons and neutrons, so theories of quarks and superstrings are erected, modified, and possibly eventually proved to be false.

As in pure science, engineering ideas must be based on knowledge and logic. To be useful they must also take account of several other factors from which scientists can remain aloof, such as economics, production, markets and timing.

Scientific induction and deduction are rarely "wrong". Inductive thought processes that transcend the confines of deductive reasoning arrive at new truths, and deductive reasoning can expand these, so long as the thinking is correct in terms of existing knowledge and is logical. Therefore, solutions to problems in science are not derived by non-scientists. Examples of thinking that violates the rules of connection with existing knowledge and of being logical are astrology and the "theory" of the Bermuda Triangle. (Religious beliefs are in a different category, of handed-down faith).

The "principle of falsifiability" is an essential feature of a scientific theory. Any theory, to be credible, must be testable against existing knowledge and by experiment. If the theory is incorrect, either in total, or, more likely, in some particular respect, it must be possible in principle to demonstrate this. Thus, for example, the truth of Ohm's Law can be demonstrated by experiments, but Maxwell proved that different approaches have to be used to predict electrical phenomena at high frequencies. "Theories" such as astrology and the Bermuda Triangle cannot be falsified by evidence or by better theories, since it is not possible to describe the cause-and-effect relationships, and no conclusive experiments can be conducted.

Part of the scientific thinking process is the creation of "models", which are mathematical formulations of the cause and effect relationships under study. For example, Boyle's Law (pressure X volume / temperature = constant) is a model of the behaviour of large numbers of gas molecules in a container. Models are the mathematical expressions of theories, and they are accepted as being true within certain known limits. The model does not necessarily represent the true situation. Boyle's Law is an empirical relationship that simplifies and averages the individual behaviour of very large numbers of molecules. Models are often developed to describe cause-and-effect relationships that are not fully understood, such as turbulent aerodynamics and material weakening due to fatigue. Models can be very useful but it is important that their limitations are taken into account when they are used to predict the outcomes of experiments. These limitations can relate to conditions of application and to precision.

ENGINEERING THINKING

Engineers do not necessarily seek to understand basic cause-and-effect relationships. Their role is to create the causes of desirable effects, useful to mankind. Engineers and applied scientists work to change basic materials and make new ones, to shape them, and to create effects based upon the laws of science, such as radio communication, transport, and electrical power generation and distribution. Since science is the basis of all engineering, engineers are taught primarily about science, and to think as scientists do.

Engineering, being the application of science, also proceeds by inductive and deductive thinking. Inventions that lead to new products or processes are usually the results of induction: examples are James Watt's piston-driven steam engine, Schottky's invention of the transistor, the Sony Walkman and Dolby's noise-reduction circuit. Deductive thinking was used to refine all of these products and to develop variations of them.

All of these products of inductive thinking were based on existing knowledge, technology, and capabilities, so they were realisable. Sometimes, however, inventions are conceived that are not practicable at the time. For example, Babbage's difference engine, a mechanical calculator invented and designed in 1834, could not work because machining capabilities could not create the precision and very low friction of the large number of rotating and sliding components involved. Leonardo da Vinci's sketch for a helicopter, using a manually driven screw-type rotor, could not work because the properties of the atmosphere meant that far more power was required than a human could provide. The idea is feasible underwater, but not in the atmosphere. Therefore practicality is an additional necessary condition for useful inductive thinking in engineering. This is analogous to the principle of falsifiability of scientific theories.

The invention of new types of electronic circuit provides good examples of the differences between inductive and deductive thinking in engineering. The separate components for a Dolby circuit all existed before the idea. The idea was how to use these in a new topology to solve the problem of noise on magnetic tape recordings, by cancellation. Deduction was then necessary to calculate the correct parameter values. Only a particular human brain possessed the combination of knowledge and inspiration to create the idea. However, any competent electronic engineer, having been taught the principles, can now design a Dolby circuit. Furthermore, circuit analysis software can be used to deduce the performance parameters of such circuit designs.

As in scientific work, most engineering is deductive. The gradual improvement of product ideas, refinements in accuracy and precision and developments in production methods are derived mainly by deductive thinking by the people involved. Many of them derive inductive ideas on the way: most engineers can remember their good ideas that occurred spontaneously. Induction comes also in small inspirations, not only in great revelations. Relatively minor ideas for improvement often generate revolutionary products. A good example is the Shimano gear system for bicycles: the world had become accustomed to the Derailleur system, which does not have an indexing mechanism, so the user must "feel" for the correct speed change lever position Shimano introduced the indexing gear system, which quickly became the new standard.

Generally, inductive developments are patentable. It is necessary to show that the idea is new, not just an improvement on an existing patent.

As in pure science, engineering ideas must be based on knowledge and logic. To be useful they must also take account of several other factors from which scientists can remain aloof, such as economics, production, markets and timing.

Engineers also use models, both those developed by scientists and others they develop themselves. Those that are accepted by mainstream science, such as Ohm's, Newton's and Boyle's Laws may be used with confidence. However, engineers also use models for processes which we create, such as those used in computer-aided design software. These are very useful but their applicability nearly always has limitations, and it is important that we are aware of these.

Models are nearly always based on the simplest explanation that describes the cause-and-effect relationships. This principle is called "Occam's razor". We are taught to seek the simplest explanations in the way that problems are set for instruction and examination. These require us to apply basic scientific laws and other appropriate scientific and mathematical concepts to solve set problems. However, real engineering problems are often not as simple as implied by these applications. Academic problems on mechanism design do not include effects such as stiction (static friction), wear, or resonance. Problems to illustrate or test knowledge of electronic principles do not include the effects of likely parameter variations or interactions.

SCEPTICISM

Scepticism is an aspect of thinking that can be constructive or destructive in the development of conceptual or progressive ideas. Obviously it is necessary to point out when ideas contravene the basic criteria of being based on knowledge and logic. In engineering we must also point out problems related to other practical aspects. The person who creates the idea must be prepared to take account of these criticisms, and if necessary to refine the idea and re-present it.

Scepticism must be based on the same criteria as inductive and deductive thinking. In fact scepticism can be considered as an aspect of deduction, the testing of the idea against the criteria. Scepticism is destructive when it is based upon ignorance, illogic, or personal motives that are at variance with the idea. For example, an idea might be adversely criticised because it would create problems for another person, or it might be viewed with envy. It can be difficult to distinguish between rational and irrational scepticism.

CREATIVITY IN NATURE AND ENGINEERING

The products of engineering endeavour are the creations of minds and hands, usually with the support of processes which are themselves selected, developed, and controlled by people. The products must be innovative, excellent, and competitive in performance, timing and price in relation to competing products. All of these qualities must be generated by the people involved, who must work together to ensure that the concept becomes a competitive product.

"Competitive products" are also developed among living species, by the operation of evolutionary pressure. As in competition between the products of engineers, the species which is best adapted to the environment and to the competition from other species will best survive. Species adapt to changes in these pressures by natural selection Over exceedingly large numbers of generations species become perfected, as a result of the process of slight random variation and survival of those that carry the best adapted genes. Natural selection has been the creator of the myriads of incredibly complex but perfectly adapted species. Structures such as single cells, insect brains and human hands are enormously more complex and refined than any human artifact. Even more complex structures, such as complete animals and, the ultimate triumph, the human brain, are vastly complex and so little comprehended that they cannot even be described in the ways that scientists and engineers explain structure and function.

Nature's creative process can be speeded up along narrow evolutionary paths by artificial means, as we observe in the selective breeding of plants and animals. However, whether measured over the millennia of natural selection or the years of selective breeding, the process is extremely slow. It is also essentially passive. Natural or forced selection cannot innovate. Therefore it cannot cope with a change that wipes out a species, and it cannot create a new species in a single generation. We cannot (yet) create a radically new species by "genetic engineering", for example, for a human to have the eyes of a cat, perfectly interfaced with the brain and other related systems such as muscles, tear ducts, and skin.

Nevertheless, nature shows how amazing levels of complexity and perfection can be attained by a process which is essentially deductive and based on very small incremental improvements, albeit passive rather than directed. Herein lies an important message for engineers: large numbers of small improvements, which can be actively pursued and implemented at will, can be much more effective than a few large ones.

Natural evolution or living species works at the level of individual atoms and molecules. The genetic code determines how these are incorporated into cells and how cells multiply and diversify to copy, with slight variations, the image of the parents. We can work with the chemical elements only at the level of very large assemblies of atoms and molecules, but we cannot yet, in any useful or practical way, manipulate individual atoms and molecules. For example, we can use an ion beam to implant phosphorus ions into a silicon wafer, and the process can be controlled in terms of depth and concentration of implantation, but it is a simple, crude, operation compared with the ability of any living structure to extract almost any element in the periodic table from its food input, and to place individual atoms exactly where required anywhere in the body.

Mankind's great advantage over nature is that, because of the unique capability of the human brain to comprehend the future, we can innovate. It is interesting to note that this ability, unique among living creatures, has been acquired only in the last 10 000 years or so of evolutionary time, when the human brain began to comprehend the concept of future time. Because of this gift, our ability to work at the macro level of materials has enabled us to create structures and materials, such as wheels, metal alloys and electronic components, which nature cannot match. They are laughably simple by the standards of the "blind watchmaker", as Richard Dawkins described the evolutionary process, but with them we can change the world, not just react to it. Our capability to innovate means that we can create wholly new artifacts, or adapt existing ones, at will and in very short times. We can create both revolutionary and evolutionary change, with the objective of creating products perfectly adapted to the preferences and constraints of their markets. The only limitations are those imposed by the laws of physics and by the extent of our imagination and ingenuity.

Creative engineering must be based upon the application of our knowledge of the scientific principles that govern the universe. There are lessons to be learned from the way that the evolution of natural species has led to the diversity and refinement of living things. However, engineering is both evolutionary and revolutionary, deductive and inductive. The present states of development of internal combustion engines and of integrated circuits are examples of evolutionary change brought about by progressive development of knowledge and improved control of variation, and they are analogous to natural creation. The invention of the internal combustion engine and the transistor are examples of revolutions, which can be created only by human ingenuity.

DETERMINISM

Applying mathematical principles to problems of science and engineering enables us to predict cause-and-effect relationships, select dimensions and parameter values, and optimize designs. Scientists take for granted the determinism provided by mathematics, and scientific theory is often derived directly by mathematical deduction. It is a scientific fact that force equals mass multiplied by acceleration, and that the electric current in a conductor is proportional to the potential difference along it, the constant of proportionality being called resistance. Such relationships enable us to turn ideas into useful products. In many cases we can safely use the relationships without adjustment or uncertainty to predict the effects of parameters and changes. Scientific and engineering education teaches that the relationships can be used, and examinations test students' ability to apply them. Obvious limitations to the theories are taught, for example no electronic engineer would apply Ohm's Law to a high-frequency problem, and mechanical engineers know that Hooke's Law applies only up to the elastic limit of the material. However, this education generates a deceptive level of credence and trust in the methods. Many real engineering problems involve combinations of effects, non-linear effects, discontinuities, and uncertainties that can lead to errors and surprises. Examples of processes that defy deterministic, linear mathematical treatment are turbulent flow, crack growth, buckling of plates and columns, stiction, insulator breakdown and data integrity. Empirical formulae and test data are often used in these situations. Other design problems involve complex combinations of known physical processes, so that the overall effect is difficult to predict or optimize. For example, a high frequency or high gain amplifier can be designed using the known component parameters and relationships, but the performance of the circuit as manufactured could be critically affected by "parasitic" parameters such as the inductance of a resistor or the mutual capacitance and inductance of neighbouring components or conductors. The life of a mechanical drive could be seriously affected by a resonant vibration caused by an out-of-balance component or a slight misalignment. As we attempt to obtain higher performance from products and systems such problems become more frequent and have more severe consequences. They can be avoided or minimized only by knowledge. Engineering is replete with traps and snares that prevent us from attaining the performance and consistency that theory indicates to be attainable.

Not all problems in engineering can be quantified in ways that are realistic or helpful. Lord Kelvin wrote:

"When you can measure what you are speaking about and express it in numbers, you know something about it; when you cannot measure it, when you cannot express it in numbers, your knowledge is of a meagre and unsatisfactory kind".

Of course, the knowledge we need in engineering is often of a "meagre and unsatisfactory kind". This does not mean that we must avoid quantification of such information or beliefs. However, it is essential that we take full account of the extent of uncertainty entailed. Kelvin's aphorism has led many engineers and managers into uncritical and dubious quantifications, which then often appear in analyses of costs, benefits, risks, or other parameters. Analyses and "models" that are based on dubious inputs, particularly those that pretend to levels of precision incompatible with the level of understanding of causes and effects, are common manifestations of the "garbage in, garbage out" principle.

There are situations in which attempts at quantification, in a predictive sense, can actually confuse and mislead. This is particularly the case in problems of quality, reliability and safety. For example, the yield of a process, such as electronic or mechanical assembly, is influenced by many factors, including component and process tolerances, machine settings, operator performance and measurement variability, any of which can make large differences to the process output. Any prediction of future yield or reliability based on past performance or on empirical evidence is subject to very large uncertainty: the yield or reliability might suddenly become zero, or be greatly reduced, because of one change. When this is corrected, yield or reliability might improve slightly, or by a large amount, depending on what other factors are involved. Yield, reliability, and factors that are dependent on them, can be predicted only for processes that are known to be fully under control. "Fully under control" is a condition that is rare in engineering, though of course we often strive to attain it. However, the more tightly a process is controlled, the greater is the divergence if perturbations occur and are not detected and corrected. In cases such as these, predictions should be based on intentions, and not merely on empirical models or past data, which can give a spurious and misleading impression that causes and effects are all understood and controlled.

There are other cases where small changes cause divergent behaviour. The final position of snooker balls after several impacts is a good example. Another is weather patterns, or the instantaneous pressure on a surface in a turbulent flow. These situations are chaotic: it is not possible to predict their exact patterns or values, even if the initial patterns and values are known, since these cannot be known with absolute accuracy.

VARIATION

There is a further aspect of uncertainty that affects engineering design and manufacture, but which is not a problem in science. Whilst theories explain cause-and-effect relationships, in practical situations we often do not know the exact values of the parameters of our designs or of the conditions that products will endure. Dimensions are always subject to tolerances. Electronic component parameter values have tolerances and can also vary with applied stress, particularly temperature, and over time. The design calculations for a product might indicate that the specification will be attained assuming all parameters are at their nominal values. We might go further, and evaluate the performance when critical parameters are at the limits of their tolerances. In many cases this deterministic approach is sufficient. However, if several parameters can affect performance, or if there are interactions between their effects, the evaluation becomes more difficult, and maybe even intractable using deterministic methods.

Production processes are always variable. The processes of manufacture, assembly, test, and measurement used to make and verify the product will all be variable. The extent and nature of the variations will in turn be variable, and often uncertain. For example, resistor values might be specified to be within ±5%, but a batch might arrive wrongly marked and be of a different tolerance, or even a different nominal value, and a machining process might be subject to a cyclical disturbance due to temperature change.

Variation exists in the behaviour and performance of people, such as machine and process operators, as well as in the machines themselves. Variation in human behaviour can be particularly difficult to understand, predict, and control, but might be the critical factor in a production operation.

Whereas variation that occurs in nature is passive and random, and improvement is based entirely on selection, the variations that affect engineering can be systematic, and we must control and minimize them by active means. Understanding, prediction, and control requires knowledge of the methods of statistics, as well as of the nature of the processes and of the people involved.

When an effect is caused by a large number of separate events or variables, it becomes safer to use empirical methods. A well-known manifestation of this is what is known in statistics as the "central limit theorem", which states that if a number of separate sources of variation combine to generate an overall variation, then the overall variation tends to be Gaussian (or "normal"). This explains the Gaussian distributions of heights of individuals and of dimensions of components machined by "in control" processes. Another manifestation is Boyle's Law, relating the pressure of a gas in a container to temperature. In fact, pressure is the overall effect of very large numbers of separate molecular collisions, of different energies, with the container walls. Because the numbers concerned are so high, the overall variation is virtually zero, and the value we call pressure can be accurately predicted.

Statistical methods are used to analyze the nature, causes, and effects of variation. However, most statistical teaching covers only rather idealized situations, seldom typical of real engineering problems. Also, many engineers receive little or no training in statistics, or they are taught statistics without engineering applications. Since variation is so crucial to much modem product and process design and to their optimization and control, this lack of understanding and application can be a severe inhibitor to progress. Conversely, proper understanding and application can lead to impressive gains in performance and productivity.

The subject of variation in engineering is covered in detail in Chapter 10.

MATERIALS, COMPONENTS, AND PROCESSES

Most engineered products are based upon the application of a range of materials, components, and processes, all of which will have been developed by other engineering teams. These developments flow continuously, so that the possibilities for products based on them increase in step. For example, the development of signal processing integrated circuits has made possible the rapid growth of a huge market for low-cost telecommunications products such as mobile telephones, and new plastic, composite, and ceramic materials have led to improvements in automobiles, aircraft, and domestic products such as cookers and kettles. A host of other components continues to be invented and improved, from fasteners and seals to microwave and digital electronic components. Newly developed processes include the robotic assembly of cars and electronic equipment, plasma treatment of metal surfaces to improve strength and corrosion resistance, and superplastic forming of metals. The scope of materials, components, and processes is so large that it cannot be covered in any but the most cursory detail in engineering curricula, and the continual developments make it very difficult for engineers to keep abreast of those appropriate to their work. However, it is as necessary to be knowledgeable in such changing aspects of technology as it is to maintain knowledge of the unchanging scientific and mathematical fundamentals.

COMPUTING POWER

Digital computers are now ubiquitous in engineering. Computers are used to design and optimize products and to control manufacturing and business processes. They are used for testing and for analyzing measurements. They form integral parts of a vast range of modem products. Engineers must use computing power for design, development, manufacture, maintenance and management. They must also know how to design computers into products. Proficiency in the use of computer aids to engineering, the design of operating and test software, and use of business software are therefore aspects of the art of modern engineering.

The use of computers in engineering has been developing so quickly, with no sign of slowing, that many engineers are unable to exploit it fully. It is very difficult to respond to the challenges and to organize systems and training that enable engineers to make best use of computers. The rapid growth in technology leads to systems having to be frequently updated, with resultant organizational and training problems, and products are quickly made obsolescent.

COST AND COMPETITIVENESS

Since the product must fill a need, and will often be in competition with other products, it must be designed and made to be attractive. Attractiveness is a combination of many factors, such as appearance, relative performance, reliability, cost, market and financial conditions, and advertising. Costs depend upon costs of materials, components, and manufacturing processes, and these also affect the performance that can be achieved. Therefore engineers must have knowledge of all of these aspects, and insufficient attention to any of them could lead to failure of the product.

TIMING

A product idea is born as a result of invention, market study, or exploitation of a new process. Initial design studies are performed, including performance and parametric calculations. Different options might be studied and compared. Eventually a particular option or group of options is chosen, and more detailed design proceeds. This includes further analysis of parameters and tolerances and optimization of the design. The manufacturing processes are also considered, and these are also designed and optimized. Materials and components are selected and operating and test software might be developed. Prototypes are made and tested, the tests covering performance, safety, durability, reliability, and other features. Eventually production begins, and the product enters the market. It must then be supported, and possibly improved, over its time in use.

This process can take a few months for a simple product, or several years for a major system like a new car or aircraft. It is a long process, but in most cases the time to market must be kept as short as possible, since the competitive pressure and commercial prospects are greatly in favour of the leading products. However, it is equally essential that the product is attractive to the market when it is launched. It has to be "right first time", and deficiencies in reliability, safety, or performance could quickly and seriously damage its reputation and sales. Engineering development must therefore proceed quickly but thoroughly, and with risks and problems anticipated and planned for as far as practicable.

The sequence of engineering activity calls for different skills as the project progresses. At the conceptual stage, inventiveness and analytical skills are needed. Depending on the technologies involved and the complexity of the design, scientific and mathematical support might also be necessary. As the project moves into more detailed design, there will be more need for engineers with knowledge of components, materials, and processes, and capable of performing detail design and test. The team might need to include engineers of different disciplines, such as electronic, pneumatic, mechanical, software, and others. They must all work together, and interfaces where the work of different engineers or teams meet must be clear and understood. In production and in service, the analytical, scientific, mathematical, and design skills are replaced by knowledge of production methods and maintenance.

Blending these skills, when some engineers will hold several whilst others are specialists in particular areas, through the development and production of the product is the art of engineering management. Managing engineering projects is intrinsically more complex than most other management tasks, because it is not possible to plan such overlapping, multi-skilled activities with confidence that they will all be completed in the planned times. Therefore planning must be flexible. Techniques such as PERT networks often cause headaches on engineering projects because of the requirement to forecast the duration of each activity. Since each engineering project is different, in terms of risks and the people and technologies involved, and since the time taken to perform knowledge-based tasks is so variable and so dependent upon unquantifiable factors, such detailed planning can be counter-productive.

It is not helpful to set deadlines and make detailed plans for scientific research. Discoveries and understanding come from patient work and thought. Because engineering is the application of science, the same applies, but to varying degrees depending on the type of product and the risk involved. However, engineering development must be driven by deadlines, and so must be planned. Finding the right balance between planning and controlling on one hand, and allowing freedom within an overall deadline, is an important aspect of the art of managing engineering. We will cover this in detail in later chapters.

CONCLUSIONS

Science is difficult. Scientific work requires intelligence, knowledge, powers of induction and deduction, and patient effort. Engineering is even more difficult. There are greater problems in terms of resources and time. Variations in human performance and design parameters make outcomes more uncertain and difficult to predict. Aspects such as production, market appeal, competition and maintenance must be considered. Several technologies might be involved, and the development teams must be multi-disciplined. Technologies, in materials, components, and methods, are continually changing, and the team must keep abreast of these. Engineers must be aware of the scientific principles and mathematical methods that are the foundations of their work. They must also be aware of the limitations of applying these to the real world of engineering, which involves variation, uncertainty, and people. Naive application of basic principles is the mark of the novice. Appreciating the complexities of the real world is the mark of experience. To create successful new products engineers must leaven the application of new ideas with experience. It is often the application of new ideas, regardless of how simple they might at first appear, that causes the greatest problems in engineering development.

However, in spite of these difficulties, or maybe more accurately because of them, engineering is often performed very well. People respond to challenges, and we see the results in the remarkable new products that flow from engineering teams, particularly in competitive markets.

The principles of management, on the other hand, are basically very simple and unvarying. No difficult theories are involved. Perversely, management is often performed very badly, or at least in ways that fall far short of releasing the full power of people and the teams they form. Sub-optimal management is commonplace in engineering.

Successful engineering depends holistically on the blend of

scientific application, empirical and mathematical optimization, inventiveness and design of the product and of its manufacturing processes, and the leadership of the project. There is no other field of human enterprise that requires such a wide range of skills, skills that can be developed and maintained only by a combination of education and continual training and experience.

Managers of creative people need to understand and abide by the simple principles in the performance of their difficult task of leadership. Carl von Clausewitz, writing in his classic book *"On War"*, stated: *"the principles of war are very simple. Wars are lost by those who forget them"*. The principles of management of engineering are equally simple, and stand like signposts in a maze of conflicting and misleading theories. Engineering managers must follow the route that has been marked. The chapters that follow attempt to show the way.

3

PEOPLE AT WORK

The traditional methods of managing people at work are based on the rational, structural, scientific thinking that is inherent in Western culture, and particularly in our scientific and engineering education. However, whilst such approaches are obviously appropriate when dealing with rational, structured systems that can be explained in scientific terms, they do not provide the best framework for dealing with human behaviour. Human behaviour is not entirely rational, and human performance is not predictable in the ways that cause-and-effect relationships in science are. The "social sciences" and "scientific management" (described below) are, at least in part, attempts to apply the principles that enable us to understand the worlds of physics, chemistry and engineering to the behaviour of people.

In the early years of the twentieth century the American engineer Frederick W. Taylor developed the principles of what he called "scientific management" [2]. In this approach, the management of industrial production processes (Taylor studied production primarily) should be based on scientific principles of measurement and control. People could be made to produce more quickly if conditions were made conducive, for example by good lighting and comfortable work positions. Working people could be helped to develop skills, and these skills are more effectively developed if the workers are made to specialize. Most importantly, the selection of workers, their training, and the methods they should use must be determined by managers, who are the "scientists" of the production process. Taylor's theory led to the development of the production line, in which workers at each stage perform tasks exactly as determined by management. "Work study" was a further development, in which tasks were broken down into basic activities such as lifting, reaching and turning, all with standard times, so that the time for each process could be minimized. Production people were not consulted, and were definitely required not to interfere with or change the processes. Management, including the immediate supervisors, imposed discipline to ensure that the methods were followed.

The Taylor principles were readily accepted by engineers, since engineers had received scientific training. What could be more natural than the application of science to their work? Production processes were simple, and production workers were mostly unskilled. The disciplined, organized approach to work also fitted the ethos of the period: people were used to the idea of being told what to do at work, and large numbers had been exposed to the unquestioning discipline of military service. Managers could conceive of no other way of establishing order and efficiency at work: any alternative would have appeared to be anarchic and unproductive.

Taylor's theories, like those of Karl Marx, contained an essential flaw. People are not amenable to scientific treatment. The factors affecting their performance are much more complex and subtle than the "scientific" ideas indicated. Individuals and groups can behave or respond in ways that are not predictable, and prediction is the logical purpose of scientific theory. This is not to say that scientific method should not be applied to work. Aspects such as ergonomics and the practical application of work study should be considered. The error in scientific management was to treat people as objects to be studied and controlled.   People at work, denied the power to influence management within the system, organized themselves, or were organized by their trade unions, in opposition to management, so that improvements in processes and productivity, for example the introduction of automation to speed production or to save money, had to be fought for.

Late in life Taylor acknowledged the flaw in his theory, and regretted the effects.

MODERN IDEAS OF MOTIVATION AND MANAGEMENT

The Hawthorne Experiments

In the early 1920s Elton Mayo conducted a series of experiments at the Hawthorne works of the General Electric Company in Chicago. These indicated that production-line workers became more productive if their working conditions were improved, for example by improving the lighting. However, production also improved when conditions were restored to the original situation, and even a control group, whose conditions were not changed, generated higher output. These results showed, for the first time in a controlled industrial experiment, that people's productivity was a complex function of individual and team dynamics, not a simple matter related to physical conditions and manual skills. A further very important conclusion was that productivity went up when the workers, as individuals and within the team, were allowed to use their own initiative to determine exactly how to perform their task of assembling telephone relays, rather than having to follow a rigid sequence laid down by management. The more intelligent workers developed better methods, by being allowed to vary them and to adapt them to their individual preferences and skills.

The subjects of the Hawthorne experiments were production-line assembly workers, whose productivity could be measured easily and unambiguously. Therefore the experiment could be conducted on conventional scientific lines, by changing the conditions and observing and measuring the changes in output, with a control experiment to ensure that the measured change was caused by the controlled variables. Even so, the results were surprising at the time, because they showed that people at work did not respond mechanically, but that their response was strongly affected by factors that cannot be quantified, such as the knowledge that they were participants in an interesting experiment, and that their advice and support were valued. Furthermore, the effects of these factors exceeded the effects of those that were measurable, such as the length of rest periods or the lighting intensity.

The Hawthorne experiments were applied to production-line workers performing repetitive, manual tasks. Most engineering work is non-repetitive, and requires high levels of mental effort. Also, the output at Hawthorne was the simple sum of the individual outputs, as the workers all performed identical, non-interacting tasks. The work of an engineering development team is highly interactive and the output is not easily measured or compared. Tasks vary in difficulty and in originality. There is often little opportunity for learning effects to generate reductions in task times, because few tasks are repetitive, and the cycle between development projects is far greater than production times for individual items. It is no doubt that for these reasons no one has attempted to emulate Elton Mayo's experiment in an engineering development environment. However, if such an experiment could be performed, there can be no doubt what the results would be.

In fact, there is considerable evidence of great improvements in productivity and inventiveness of engineering teams that have been allowed freedom from the constraints of organization and procedures. We will consider examples of these later.

Maslow's "Hierarchy of Needs"

In 1943 A.H. Maslow postulated the "hierarchy of needs" (later described in his book *Motivation and Personality* [3]) as the basic motivating forces of people in life and at work. The hierarchy of needs is:

1. Physical: freedom from hunger, thirst, and pain.
2. Security: absence of fear.
3. Social: the need to feel accepted and wanted by neighbours and people at work.
4. Esteem: the need for personal recognition and respect.
5. Self-actualization: the need for personal fulfillment and realization of personal potential.

Maslow reasoned that people will be motivated by the most pressing unsatisfied needs in the hierarchy, and that higher needs will not motivate people who had not satisfied lower needs. For example, a person who is grossly underpaid or under threat of redundancy will not be too concerned about whether the value of his work is appreciated, and therefore an indication of such appreciation will not improve his performance. In fact, it might even be counter-productive, since it is likely to be received with some scepticism. Whilst the higher level needs are very powerful motivators, they become switched on only when the lower level needs have been satisfied.

Mcgregor's Theories of Management

In 1960 D.M. Mcgregor described what he called the 'Theory X" and 'Theory Y" management styles (*The Human Side of Enterprise* [4]). 'Theory X" managers are authoritarian, aloof, and manage by orders and decrees. They do not tolerate deviations. They believe that they, and lower managers appointed by them, should decide how tasks should be performed, and that the people working for them are not competent to determine working methods themselves. This style of management fits well with Taylor's theory of "scientific" management, and it satisfies the first two levels of Maslow's hierarchy of needs. '"Theory X" managers can be effective, particularly in short-term situations or in crises, or when the tasks to be performed are simple. Of course very few engineering tasks fall into these categories.

"Theory Y" managers, on the other hand, adopt a relaxed approach. They are approachable, and manage by consultation and consensus.

They minimize the use of directives and formal procedures. "Theory Y" management is based upon the following assumptions:

1. People do not inherently dislike work, and in fact prefer to enjoy it.
2. People have a natural tendency to direct and control their own life and work
3. People's commitment to work is related to the rewards they receive, including recognition, respect, and fulfillment (Maslow's 'higher needs").
4. People can learn to accept responsibility.
5. Most people want to apply imagination and ingenuity to their lives and work.
6. Industrial managers generally underestimate the intellectual potential of people at work.

Mcgregor explained why "'Theory Y" managers are usually more successful in terms of the results of their teams, particularly in the longer term and when the skills and experience of individuals in the team are important in achieving the objectives.

In most management situations there is a constant tension between the two approaches, which is influenced by personalities and priorities. Personalities are reflected in the style adopted: it is often the under-confident or less competent manager who will resort to the "Theory X" style. If short-term success is achieved, the manager will reap the reward of personal recognition, but he is unlikely to have gained much admiration from his subordinates, and long-term consequences for the business might be adverse, with other managers having to cope with the damage. Senior managers often put "hatchet men" in place to sort out project problems. The "hatchet men" are promoted as a result, and then spread the "Theory X" culture more widely.

The overriding importance of developing and maintaining the "'Theory Y" style of management, particularly in engineering, will be discussed later in this chapter.

Drucker's "New Management"

Drucker, in his book *The Practice of Management* [1], explained why "scientific management" is fundamentally inappropriate to the motivation and management of people at work. He taught that people at work want and need to feel that they can contribute to the enterprise, at whatever their level of operation. Also, the fact that they are trained and experienced makes them competent to do so. Furthermore, such contribution must be demanded of them. All workers are managers, and lower level workers have knowledge that higher managers often do not. By empowering and requiring them to apply that knowledge, management can generate greater gains than if they plan and control the details. No "scientific" principles are involved. Instead, the facts of human nature are recognized and turned to the benefit of the enterprise. All people in the enterprise, from the chief executive to the lowest level, are both managers and workers, and are on the same team.

Drucker therefore advised that responsibilities should be delegated to the people who perform the tasks, and that management's role is to determine the major objectives of the business and to set up the necessary structure and resources, and then to motivate and train their people to perform in the right direction.

Drucker extended the argument to organizational aspects. If responsibilities are to be delegated and people trusted, then the organization should reflect this. Project managers should be given responsibility and authority and the freedom to make decisions within broad directives. In military terms, they should determine tactics, within the overall management strategy.

Fundamental to Drucker's teaching is the imperative that teams should be created, working for the managers responsible for meeting the objectives. The managers must be responsible for leading their teams in all respects that affect the performance of their work. Outside interference from specialists or "coordinators" should be eliminated. The managers must of course report progress and problems to higher management.

Managers are the most expensive resource of the enterprise, and also the resource that most influences both short-term and long-term performance, including profitability, growth, and survival. The quality of management is the only factor that really differentiates competing businesses, since they all have the same access to machines, labour and other resources that can be bought. Management determines how well the resources are applied, and it is the application of the resources of business that determines success or failure.

Managers, acting as leaders, as planners who must always balance short-term and long-term requirements and objectives, and as educators and inspirers, cannot be created merely by promoting people into "management" grades. Rather, managerial talent must be identified by results and character, then stimulated and improved by training and experience. Managers must show integrity and moral responsibility, and must have a sense of vision. They must be teachers so that they can develop their subordinates. These characteristics cannot be learned or taught, though they can be improved by suitable training and motivation, so wise selection of managers is crucial to the performance and future of the enterprise.

Drucker's teaching is at the heart of all successful modem business.* Indeed, Drucker predicted that the countries that would be the world's economic leaders in the second half of the twentieth century would be those whose managers understood and practised the principles he set out. Japanese industry systematically applied these principles after 1955, as did most of the companies in other countries that have become world leaders in their fields. The converse is also the case: the countries with politico-economic systems that have prevented the development of the new management have been unable to compete, leading to the collapse of the entire system, again as predicted by Drucker. Industries and companies in the Western democracies that have not embraced the new management principles have lost ground, and in several cases have been driven to extinction by their more effective competitors. These contrasts exist starkly within the engineering industries.

Deming's 14 Points

W.E. Deming is best known for his "14 points for management" which he described in his book *Quality, Productivity and Competitive Position* in 1982. The book was republished under the title *Out of the Crisis* in 1986 [5]. Deming had been teaching methods for improvement of quality and productivity in Japan since 1950, having been invited to assist Japanese industry to rebuild after the war. With W.A. Shewhart, he had been one of the pioneers in teaching the application of statistics to production quality control in the United States during and shortly after the war. His teaching was widely accepted in Japan (with that of others, notably Kaoru Ishikawa), and he is now acknowledged to have been probably the most influential single figure in the growth of Japanese industry since 1950. The annual Japanese award for excellent progress in quality is called the Deming Award. Deming was not "discovered" in the United States until 1980,

---

* Peter Drucker was born in 1909, and in 2004 is still teaching and writing. His other notable books include *Managing in Turbulent Times* and *The Effective Executive*. Visit www.peter-drucker.com.

when a TV programme entitled *"If Japan can, why can't we?"* brought his name to wide public attention, and since then he has been in great demand as a consultant and lecturer, and national "Deming Associations" have been formed in several countries. The 14 points are:

1. Create constancy of purpose for continual improvement of products and service, allocating resources to provide for long-range needs rather than only short-term profitability.
2. Adopt wholeheartedly the philosophy of continuous improvement and elimination of all waste, delays, and defects.
3. Cease dependence on inspection to assure quality, by creating quality products in the first place.
4. End the practice of awarding business on the basis of initial cost only, and insist upon quality. Develop partnership arrangements with single suppliers rather than adversarial dealings with multiple suppliers.
5. Strive to improve continuously every process of production, planning, and service.
6. Institute continuous on-the-job training for all, including managers, to develop new skills and to keep up to date with new methods, materials, etc.
7. Provide leadership at all levels of management, to help people to do a better job.
8. Eliminate fear from the workplace by encouraging effective communication between all in the organization.
9. Break down the barriers that usually exist between departments and functions, by instituting teamwork
10. Eliminate slogans, exhortations, and targets that demand higher productivity without providing methods.
11. Eliminate work standards and numerical quotas. Replace them with help and leadership towards continual improvement.
12. Remove all barriers that prevent people from having pride in their work.
13. Institute a comprehensive programme of education, and encourage self-improvement.
14. Ensure that top management is totally and permanently committed to continuous improvement in quality and productivity, and that the commitment is shared at all levels, by active leadership and participation.

Deming's 14 points clearly are in harmony with the earlier Western teaching on motivation in industry. In particular, they apply the ideas of Drucker and Mcgregor to the business of engineering production, emphasizing continual improvement through delegation and teamwork. Deming also emphasized the effective use of statistical methods as one of the essential tools for improvement.

The unifying principle of the work of people such as Mayo, Maslow, Mcgregor, Drucker, and Deming is that people want to excel as individuals. They need to feel that the value of their work is appreciated. Therefore they will respond to efforts to help them to perform better. However, they must be confident that such efforts do not put at risk their other, more basic needs, particularly, in the industrial situation, for job security. When the output is that of a team, the individuals want the team to excel, so the team aims must be harmonized with the aims of the individuals. It is ironic that the great transformation of Japanese competitiveness was instituted by Americans, teaching methods that had been taught but not adopted in the West. The industrial transformation took place mainly in engineering, particularly in the mass production of vehicles, electronic components and consumer goods, though of course the message applies to all types of human endeavour.

Later Writers

Some more recent writers have added valuable contributions. Peter Senge, in *The Fifth Discipline: the Art and Practice of the Learning Organization* [6], explained why scientific reductionism, inherent in scientific management, is the wrong approach for dealing with the complexity and change that characterizes most modern business. The "fifth discipline" is "systems thinking", the ability to view and understand the totality of the business and the environment in which it must survive. A key aspect of this must be continuous learning by everyone concerned with managing any aspect of the business.

Tom Peters and Robert Waterman (*In Search of Excellence*) [7], Rosabeth Kanter (*The Change Masters*) [8] and Gary Hamel and C.K. Pralahad (*Competing for the Future*) [9] make interesting and important contributions, consistent with Drucker's teaching. They all emphasize the need for businesses to be agile and responsive to changes in technologies and markets.

Brian Thomas' *The Human Dimension of Quality* [10] emphasizes that all success is derived from the people involved, and that a balance must be struck between the kind of aggressive change advocated by some management teachers and the need to take account of how people respond to change. He makes interesting comparisons between the development and survival of organizations and of natural species. Thomas is not as well known as the writers discussed above, but his book is a most valuable contribution.

K. Murata's excellent book *How to Make Japanese Management Methods Work in the West* [11] describes his experiences in setting up and running a manufacturing plant in the UK. He provides simple but elegant descriptions of the methods that led to success, all consistent with the new management.

INDIVIDUALS' TALENTS AND MOTIVATIONS

Individuals have talents that vary over wide ranges of type and quality. Some of these are fixed within the individual. Basic intelligence and aptitudes are probably determined at birth, so that a person who is good at, say, artistic expression should not necessarily be expected to be a good engineer. Likewise, a person might be extremely inventive, but lack the aptitudes needed to manage a team. Not all engineers are inventive, inductive thinkers.

People also have other talents, which develop with time. These are the skills taught by education and training and developed with experience. Anyone can be taught skills, but responsiveness and the levels attained will depend upon innate talents, quality of teaching and motivation to learn and improve. Whilst an innate talent such as intelligence or inventiveness is limited within any individual, there are no limits to the knowledge and skills that can be developed. So long as opportunities and motivation exist, an individual's knowledge and skills can be improved continuously.

A person's motivation to improve depends strongly upon the factors described by Maslow in the "hierarchy of needs". The higher needs, for social acceptance, esteem, and fulfillment, must be seen to be satisfied by the effort to obtain further skills and qualifications. Since this effort is often considerable, there must be a correspondingly strong motivation. This can be provided by management if the organization recognizes and rewards the improvement, particularly if it helps the individual by guidance and other support, and then makes use of the additional skills to the benefit of the organization.

Individual motivation also depends on other more subtle factors, which can be more difficult for managers to appreciate and influence. Age, family situations and health can have powerful effects, either in favour of or against performance and improvement. These can vary quickly or over a long period, and there may be little or no outward signs or information. Moods and attitudes can vary due to unconnected events or to other information, notably rumours.

All of the factors that affect motivation exist in unique and varying combinations within every individual. The effects on overall motivation cannot be measured or predicted. The effects are not simple linear combinations, so that, for example, 5% more pay would not necessarily justify having to work at weekends. The effects of individual factors and combinations can be quite disproportionate, so that an apparently minor disagreement between a person and his supervisor could cause a significant reduction in productivity, despite most other factors being favourable. Conversely, a small boost to self-esteem due to the acceptance of an idea could result in a determined effort to generate further improvements, or in sub-conscious but significant smaller improvements.

The extent to which motivational factors influence productivity and self-improvement also depends upon the type of work involved. At one extreme, purely manual, unskilled work, whose output can be easily supervised and measured, will be affected, as Mayo found in the Hawthorne experiment. At the other extreme, an individual scientific researcher working alone might not need any further motivation than the interest and esteem generated by his work, and to him training programmes and visits from his department head might seem to be unnecessary distractions. However, the productivity of knowledge workers such as engineers, whose work is difficult to supervise and measure, is critically dependent upon motivational factors.

TEAMS

Whilst individuals each possess a range of talents, teams of people working together possess the sum of these talents. If we need a team that includes skills in analogue electronic design and motor control, we can either employ two specialists or one engineer who is experienced in both areas.

However, the motivation of teams is even more complex and sensitive than that of individuals. Whilst individuals have complex and changing motivations, the motivations of groups are convolutions of individual motivations. In addition to the non-linear effects of motivations on the individuals, the effects on groups is the combination of the individual non-linearities, coupled with interactions and feedback. Group response is inherently chaotic, as extremist politicians and religious fanatics have demonstrated and used. Like other chaotic systems in nature, the behaviour of groups of people can be generally predictable, but can show very large divergences in any direction. In the management context, groups can be moderately co-operative and productive, finding a nearly stable equilibrium of motivation and effort, but they can become extraordinarily non-cooperative and hostile to the aims of management. On the other hand, they can also be motivated to extraordinary levels of enthusiasm and productivity. Most of us are familiar with situations in which a group of people has been very highly motivated to achieve surprising levels of performance. This happens when the group is united in the challenge to be faced and the way to meet it. The challenge is nearly always a severe one. The other common factor in these situations is the happiness of the team: despite the challenge, which might even involve negative motivators such as fear or discomfort, the group shows joy in working together.

The performance of teams is, of course, strongly influenced by their leaders. Team leaders are imposed by management. However, in many groups unofficial leaders also emerge. These are individuals who possess qualities that attract respect from the other members, who are prepared to give up some of their independence of thought and action in favour of perceived more effective satisfaction of their demands and desires through the leader. Unofficial leaders almost always emerge when the imposed leaders are not effective, or when the group perceives that their objectives are not the same as those of the imposed leader. This is a common situation in management, where the aims of the organization are perceived to be hostile to those of the individual, for example pay vs. profits. Many less obvious disunities can occur, such as a manager who is considered to be using his group's performance as a prop for his own self-esteem and promotion.

Unofficial leaders emerge strongly where trade unions are involved in representing and protecting individuals at work. In these situations management has little option other than to recognize and work with these representatives, who will usually have a semi-official role in terms of management influence.

Of course the ideal situation is for the imposed leader to be accepted by the group. For this to happen, the aims of the leader and of the team must be perceived to be in harmony, and the leader must have the attributes that inspire the team to accept his position and to trust him to satisfy their higher motivational needs. This in turn means that they must trust top management in the same way, since the leader represents top management and is put in place to execute their objectives.

Whilst allowing freedom to select and develop methods of work maximizes initiative and performance, it is nearly always necessary to impose some constraints. An obvious one is that if people are to work as a team, their work times must be harmonized. If a design is being created using a computer-aided engineering (CAE) system, all designers must use the same system, or at least compatible systems. Standardized approaches to certain tasks, such as component selection and software documentation, can help to improve productivity by reducing "wheel re-invention" and by ensuring that tasks that are essential but which might be overlooked are performed. Standardised approaches also greatly facilitate training. Therefore, there is a minimum level of rules and procedures that does need to be applied. Just what that level is will depend on the situation, and it can never be determined exactly. Any procedure that is helpful and useful is worthwhile. However, if it is buried in a mass of less helpful procedures the whole system could be ignored, including the few good procedures.

There is a good and valid analogy between the management of teams at work and that of teams in sports. In games such as rugby and football (soccer, for Americans), rules exist which govern how the game is to be played. The rules ensure that both teams are playing the same game, and they have been developed over time to improve the quality of the game, for players and spectators. Every player knows the rules, to a high level; in fact detailed knowledge of the rules is essential for good players. Everyone plays to the rules, but if anyone does transgress, the referee will stop the game and award a penalty. All players accept the authority of the referee to apply the rules. However, the rules do not inhibit originality or individual effort within the range of freedom permitted. For example, in rugby the ball may not be thrown forward, but the player with the ball has absolute discretion how to kick and pass subject to that limitation. When a player has the ball, all other players act in support, because they all aim for the same objective, to win. The team that achieves the best blend of individual and team performance will usually win (though, as in engineering and life in general, luck also plays a part). Therefore team managers and coaches ensure that individual and team training and motivation enhance these skills. A coach might work out moves and have the team practise these, but he will not impose them as the only way to play. Finally, all players enjoy the coaching and the game, even though it involves discipline, training, effort, discomfort, and possibly pain. Of course the winners usually enjoy the game more.

Success in team games such as these illustrates the optimum application of nearly all of the principles of good modem management described earlier. "Theory X" management style cannot create teams that win. Players perform best when they are free of basic worries, are esteemed by their team mates as valuable members, and when their skills and the success of the team gives them pride and confidence. Teams perform best when all players share these motivations, and when they are managed and coached to fuse the individual talents into a united, creative force.

The team "managers" must optimize the contributions of the individual stars and those of the team working together. An uncoordinated team of individual stars will not be successful, but we often observe surprisingly good performance from excellent teams that do not contain top rank players. The "managers" must select the right players to play in particular positions, decide the strategy for individual games, and then brief the team. The management roles are dearly defined, and there is total delegation of tactical decision-making. Sports team managers know their sport. Their capabilities, as managers, coaches, or captains have been identified and developed as a result of their participation as players. Tennis players do not qualify as football coaches.

Most people find these ideas easy to accept in the context of sports teams. They would consider it perverse if a team captain instructed his team to ask his permission before each attempt to score, and allowed only certain members to do so. The team members would probably not bother, and would go and play for another team. If the rules of the game discouraged bold new tactics or moves, people would play a different game. Yet these types of situation are common in business, not least in engineering. The management of engineers and of engineering must be based on the same fundamental principles that underlie success in sport, since the resources that matter most, the people in the team, are the same.

Committees and Meetings

A committee is a special form of team. Committees are formed to deal with a defined topic over an indefinite period or for a limited time, typically until a problem is solved or a report is produced. It is fashionable to give fancy names to committees, like "executive committee", "action team" or "focus group". Committees can proliferate to an extent that becomes wasteful and demoralising. With the growth of the idea (or "paradigm") of "empowerment" it has become common for committees to be set up to deal with problems that could be better solved by individual action or by simply talking to the right people. It is common for people to sit in on committee meetings for hours, and make a contribution that takes only minutes. However, being on one confers status and self-esteem, but being left off implies that being out of touch, unimportant or superfluous. It is also interesting how long a discussion can continue long after it is clear that no one present is in disagreement.

Committees are led ("chaired") by people who exert influence or power. It is important that the selection of people for these roles is consistent with the normal management system and does not create confusion or conflict.

The creation, composition and running of committees should be carefully managed. There should be a disciplined framework for approving the formation of committees and for terminating them when their work is done. When only brief contributions are required the people concerned should be invited to attend only for the time necessary. A useful maxim is: "appoint yourself a committee of one, to see that your work is properly done".

STRESS

Stress seems to have become an inescapable feature of the modern workplace. Theory X management seems to be ascendant, "change programmes" generate uncertainty and confusion, "empowerment" as a substitute for effective help and leadership sows distrust and more confusion, people are expected to work harder and longer, "downsizing" and arbitrary culls of people create fear and job security is a distant memory. Loyalty between workers and employers, in both directions, is not expected nor extended.* Managers now have to cope with the adverse effects of stress on their performance and on their teams'. A large proportion of time taken off for health reasons is stress-related. Specialist "stress counsellors" are appointed and government departments initiate "stress awareness" campaigns.

Of course not all stress is damaging. Transient job pressures, personality issues and personal problems have always generated

---

* All of these developments run counter to the principles of the new management. Why scientific management is so resurgent and the new management so forgotten is an interesting question for our times, and we will return to this in Chapter 12. I have just (2004) reviewed three recent books on engineering management. None of them even mentions Drucker!

stress, for short periods or among a few people, but good managers have been able to mitigate the effects and deal with the causes. However, unremitting, widespread stress is a new problem, a modern industrial disease. It is not at all easy to cure, but it can be largely prevented by adherence to the new management principles.

Stress should not be a problem in well-managed engineering organizations. Nearly all engineering work is inherently creative and interesting, and engineers are intelligent, rational people. If engineers are bored their managers are not leading them effectively. The pace of work and its quality can be best assured by the sort of leadership discussed earlier, and in more detail in the next chapter. Happy people work better, and stressed people are not happy.

CONFLICT

Engineering teams must work together for long periods, and members join and leave as projects proceed and as individuals are promoted, transferred and retired. There is often a wide range of ages and backgrounds. The degree of team and individual working can vary. Inevitably there will be different possible approaches to problems, and priorities that seem to be in conflict. There are nearly always pressures of cost and schedule, and internal competition for favour and promotion. All of these factors increase the likelihood that team members will occasionally find it difficult to work harmoniously together. Combined with all of these causes for potential friction and conflict in engineering work is the fact that engineers are no different to other professions in terms of personalities, moods, and pressures that can influence teamwork.

However, effective teamwork is crucial to successful engineering, and the converse also holds: any conflict that damages teamwork can seriously impair the chances for success. Yet it is inevitable that friction and conflict will occasionally arise between individuals in teams.

Engineering managers must first seek to prevent friction and conflict by careful selection, development and leadership. They must be alert to potential conflicts and must act swiftly to prevent them from developing. They can do this only if they work closely with their teams, keeping alert to signs of trouble, listening and observing. The team members must see the manager as a colleague, not merely as an agent of higher management, or the early indications of conflicts could be hidden. Potential conflicts can often be averted by early friendly discussion, with the manager acting as arbiter. He must take action to remove the cause. Merely telling people to solve the problem between themselves will not work, and will probably worsen the situation. Rational individuals do not want to be in conflict, so if such a situation does arise it is unlikely that they will be able to resolve it between themselves. The action could be to make a decision that the antagonists and the other team members involved will accept as clearing the problem, or to reassign responsibilities.

There are two kinds of conflict within teams. Two individuals may be in disagreement, or an individual may be in disagreement with the rest of the team. In the former case the manager must talk to both, together and individually, and then must act as described above. Ideally both individuals will be satisfied with the outcome, but of course there will be times when the manager must come down on one side or the other. When the problem exists between an individual and the rest of the team it is unlikely that the individual is right and the others wrong. The manager must therefore come down on the side of the team. However, the cause of the problem must be dealt with, and it is even possible that, for example, the individual has a genuine point that needs to be explained to the rest of the team.

If conflict has already developed action must be taken immediately to remove the cause and to re-establish harmony. The team will seldom be able to solve the problem without assistance, and so will look to the manager for a solution. His effectiveness in doing so will greatly influence the respect and support he will receive in future. Most conflict situations are temporary, involving for example disagreement over workspace or use of facilities, and these can usually be patched up without long-term damage. Other problems can be more deep-seated, such as a personality clash between individuals, when more careful handling, possibly with help from higher management or from a specialist counsellor is necessary. In extreme cases people must be removed from the team, or even fired as a last resort if there is a disciplinary problem. Removing people can be upsetting to the project, especially if they are key members with skills that are difficult to replace. Sometimes their services to the project can be retained by making them contribute via a different route, such as through a function manager not working directly on the project. Action to remove people from a team is always unsettling and unpleasant. However, it must not be delayed or avoided if it is necessary. When trade union or internally agreed procedures exist for dispute resolution these must, of course, be followed explicitly.

Conflict can also occur vertically within organizations, for example between a manager and his boss, or laterally between different managers. The principles discussed above apply to these problems also: the manager at the next level up has the responsibility to solve such disputes.

Conflict can actually be beneficial, if it results in greater respect and cohesion later. This is often the result of brief flare-ups followed by apologies, as every good parent knows. If conflicts can be dealt with quickly and openly, rather than being allowed to submerge into long-term uncooperativeness or hostility, the people involved can learn more about one another and about themselves, and can be forged into a happier and more effective team. Teams and organizations that are made up of people who have been carefully selected and trained, and who are led with skill, sympathy and discipline will be most likely to work in harmony. Preventing conflict must be a prime objective of managers, at all levels. Eliminating conflict can be very difficult and can even put managers under strong emotional stress. As a result such problems are often put aside or dealt with unsympathetically. How managers solve conflict situations can greatly influence the effectiveness of teams, which can even be strengthened and made more cohesive by courageous action. It is also an excellent indicator of leadership ability.

STUPIDITY

We have all, like everyone who lives, works and otherwise interacts with other people, observed stupidity in action. Albert Einstein observed: "the Universe and stupidity are infinite. I am more sure of the infinity of stupidity than of the Universe." Stupidity is a fundamental characteristic of people: we all perpetrate it and we all suffer the consequences. As the great sage Dilbert* said, people are all greedy, horny and stupid. Other people, of course; not us. However, stupidity is a powerful force that influences events and destinies. Psychologists, sociologists and others have studied and derived ways of measuring intelligence, but few (except maybe Dilbert, but he is unreal) have studied stupidity. The various forms of madness have been researched, but madness is not the same as stupidity. Very sane people can be stupid, or do stupid things. We all do, as Dilbert observed. Some of us more than others, but no one is immune.

Stupidity is also not the opposite of intelligence. People have high, low or in-between intelligence, and we can observe it and even measure it. Stupidity is altogether another dimension of human

---

* For those unfortunate enough not to have experienced Dilbert, he is the hero of Scott Adams' excellent cartoon strip. Visit www.dilbert.com

propensity. Intelligent people can behave stupidly and people with lower intelligence can be wise. (Maybe extreme intelligence is positively correlated with stupidity?).

Since stupidity affects us all, it is more pervasive than madness or dumbness. It is more difficult to deal with, because it cannot be measured, treated or improved. It is also dynamic and changing, so that very sensible people can suddenly do or say something really stupid.

Dale Carnegie, the great teacher of human relationships and author of *How to Win Friends and Influence People*, did not deal with stupidity. All of his lessons and examples imply that the people to be influenced and won over are basically sensible and rational, and not driven by greed, lust, hatred or cowardice. These and other emotions are driven by forces like motives, experience, genes, body language, sex appeal, and propaganda. Therefore they can be anticipated and dealt with. We can persuade people to drive carefully, to walk towards machineguns or to refrain from beating their wives, by persuasion or by fear of punishment. But we cannot anticipate, manage or prevent stupidity. It just happens.

If individual people can be stupid, groups can be surpassingly so. Stupidity congregates. We know that tigers and eagles are smarter than sheep and turkeys. But even smart people can coalesce into stupid groups.

Since everyone can be stupid at times, and since groups so often behave stupidly, coping with stupidity at work is a daunting challenge.

CONCLUSIONS: APPLYING THE NEW MANAGEMENT TO ENGINEERING

Many engineers find it difficult to unshackle themselves from the ideas of "scientific" management, and many others are not even aware that there is a better way. Engineering is based on science; indeed, numeracy and rationality are essential qualities for successful engineering. Therefore it is natural for engineers to plan, organize, measure and control. Whilst these activities are obviously necessary, the new management philosophy demands that they be kept to minimum levels necessary, and not be allowed to stifle initiative and creativity. Striking the right balance and withstanding the pressures and temptations to extend control is the most difficult, yet the most essential component of the art of management. The balance will be different for different types of engineering activity: conceptual design should be subject to fewer constraints and disciplines than a production or maintenance operation. The skills and experience of the managers and the other team members also influence the balance. However, most engineering work is much more uncertain and subject to greater change and risk than most other tasks that people manage. Therefore principles of planning and control that are useful in, say, retailing or housebuilding can be severely counter-productive when applied to engineering.

Liberating and empowering engineers by giving them broad but challenging objectives, such as the main performance and cost targets and time to market, then giving them the authority to make and change their own detailed plans, generates surprising results in ingenuity and productivity. Likewise, engineers at all levels in continuous operations such as production or maintenance can generate continuous improvements in productivity and quality, which in total can greatly exceed the effects of occasional "campaigns" instituted by higher management. This empowerment can be achieved only by creating teams, and then demanding that they excel. The teams must include all of the necessary skills, at the right times. In engineering the nature and rules of the game can change, for example as products pass through different phases of development and production the skills and disciplines involved will change, so engineering teams are not usually fixed, yet must always be complete and fit to win. This approach, which many would consider to be a commonsense one, has been re-invented and given names such as "concurrent" or "simultaneous" engineering, and books are written and seminars are held to extol the idea.

The new management allows and encourages both inductive and deductive thinking to be applied to problems. "Scientific" management, or any tendency to over-plan or over-control, discourages inductive thinking. Inductively derived ideas usually run counter to procedures and upset the "system". They force people to have to think and to re-evaluate current methods, and this can be uncomfortable. Yet inductive breakthroughs, at all levels, are essential for progress and improvement in competitive situations. The total management philosophy must therefore stimulate inductive thinking and application of the ideas that result, as well as those that flow from the more continuous deductive process.

The new management philosophy places great demands of leadership, knowledge, vision, and courage, at all levels of management. All levels must have knowledge and skills to teach their people and to guide them when problems arise. They must be able to judge which methods and ideas are likely to succeed and which not. They must have the talents and personalities to develop and exploit the skills and enthusiasm of individuals and to create and lead teams that work to the objectives determined by the managers.

Management talents such as these are obviously not commonplace. On the other hand, it is the quality of management at all levels, and particularly at the top of the organization, that is the principal determinant of success and growth, especially in competitive situations. Therefore the most crucial task of senior management is to select and develop the management resources. How this should be achieved in engineering is the subject of the next chapter.

4

DEVELOPING ENGINEERS

THE ROLES OF ENGINEERS

Engineers perform two prime roles in most jobs: they must be professionals, and they must be managers. There is seldom a sharp division, since nearly all engineering jobs entail some management. Engineers working as specialists or as individuals within a team must control their schedules and budgets, and often must participate in planning and liaison with other team members, suppliers and others. At the other extreme, many engineers fulfill roles that are primarily managerial, planning and managing the work of others.

Both roles require the particular knowledge and skills of engineering, as appropriate to the product or service being provided. Obviously the engineer designing a vehicle engine fuel controller must be highly skilled and experienced in the technology concerned, otherwise his productivity and the quality of his design will be compromised. The manager of the development team for the whole engine system need not be a specialist in fuel control, but he must understand and have experience of engine systems and their development. Without this he would be unable to lead the project effectively, he would not be able to perform the many judgments necessary to balance risks, opportunities and costs, and most importantly, he would find it difficult to hold the respect of the team.

The necessity for engineering managers to be competent engineers in the disciplines they manage is absolute. It has become a widely believed myth in much of Western industry that engineers are not the best chief executives of engineering companies, or even the best leaders of engineering projects, since they are considered to have insufficient grasp of such business essentials as contracts, finance and markets. This belief is reinforced by tales of projects and businesses that have failed because their engineer managers paid inadequate attention to the commercial aspects, or underestimated the risks of the projects. Whilst these stories are nearly all true, they misrepresent the fundamental truth. In fact, the most successful engineering enterprises are usually managed by engineers (particularly in Japan**, but there are many notable Western examples, including HP, Agilent, Rolls Royce Aero Engines, GE, and probably most small and medium sized companies), and in those that are not the chief executives ensure that their engineering managers are given the responsibility and authority necessary to manage the engineering functions, as well as primacy among the managers of other functions. Misjudgments are made by all managers, and the nature of engineering decision-making in times of rapid technological and market changes and competition for markets makes such judgments prone to great uncertainty. This accentuates the need for judgments to be made by people who can understand the technological risks and opportunities, as well as the commercial aspects.

There are overriding reasons why engineers must manage engineering. Only an engineer can understand the work. Since engineering is the application of science, the knowledge necessary cannot be "picked up" by people trained in non-scientific disciplines such as accountancy or law. On the other hand, most engineers, since they must be numerate, reasonably intelligent and rational, can be educated to a level sufficient to understand the aspects of accountancy, law and other disciplines that affect the project and the business. Accountants, lawyers and other specialists are employed to support the business. They perform necessary functions and provide advice, much of which will influence engineering decisions. However, they cannot make engineering decisions, since they cannot judge the risks and manage the engineering effort. Therefore they cannot determine the future for the enterprise, and the primary role of management is to secure the future.

When engineering enterprises are run by non-engineers, or when engineers in management lose management influence, there is a strong tendency for short term expedients to take precedence over long term planning, particularly in areas that do not appear to influence direct costs. This manifests itself in the accountant's classification of training and other developmental activities as "overheads", to be reduced as much as possible, particularly when there is pressure on short term profits. In fact, as will be explained in more detail later, activities such

---

* In 1992 Akio Morita, founder and chairman of Sony Corporation, in a speech to the Royal Society in London, said "manufacturing and high technology corporations must be led by those who understand not just business but technology as well." He went on to explain why, and this section reflects his wisdom.

as training, staff improvement and research are necessary investments, to be balanced against the expenditure necessary for immediate operation in order to ensure future growth and survival. The accountant can advise on some of the effects of decisions, but only those that are easy to forecast, such as costs. He cannot judge the much more important and longer term effects, such as the benefits of having better qualified and motivated staff or the payoff from a particular line of research. These are not precisely quantifiable, but they represent present costs that are, and they are therefore perceived as easy targets for cost cutting.

Another accountancy-based policy is that of buying at lowest cost, whether of supplies or subcontract work. Lowest cost purchasing, and the related practice of seeking tenders for work and then selecting the cheapest, is probably the single most damaging idea ever developed in the running of business, particularly in engineering. We will discuss this in more detail in later chapters.

SELECTING ENGINEERS

Specialist Engineers

All engineers must do engineering, and for many it is their main role. The designers, test engineers, production engineers and maintenance engineers are professionals in their fields, and their productivity depends upon their skills and on how effectively they are managed. Though they must be expected and encouraged to manage their part of the overall task, in conjunction with the rest of the team, they would not normally spend much time on management tasks such as planning, organizing and leading. Specialist engineers are created by the educational system, which makes them available to enterprises. The first management role for the engineering enterprise is therefore to select the specialist engineers it needs. It bears stressing, however, that the enterprise must not confuse this task with selecting managers: we will cover this subject later.

Whilst managers in industry cannot control the teaching at the schools and universities that will ultimately create the engineers they need, they can and should influence both students and teachers. They can do this by maintaining contacts with local schools and with selected universities and other higher education institutes, for example by organizing visits by students to the company, by providing occasional lecturers from company staff, and by donating equipment to laboratories and workshops. Universities can be used for extramural research. Such bonds are altruistic in that they encourage young people into engineering, but they also create a favourable image for the company among jobseekers. Since the future of the company depends so much on the quality of its people, every effort should be made to ensure that the best are encouraged to join.

People enter engineering jobs at every level from school leavers to postgraduates, and with experience from nil to many years. Having determined what kinds of people are needed, managers are faced with the vital task of selection. We will deal with this at different levels, starting at the most basic.

School leavers join engineering at the apprentice or technician level*. Their early work may involve mostly initial training, followed progressively by supervised tasks. These people are not yet "professional" engineers, but one day they might be, and their selection should recognize this, particularly as they are likely to spend many more years in the company.

There are two basic criteria for selection of such young people into engineering jobs. The first is their basic aptitudes, which of course must include mathematics and science, particularly physics, as well as the ability to write. These aptitudes can be determined easily from inspection of school reports and examination grades. The second is attitude. Attitude is important, since engineering work and the development of people in engineering involves teamwork, willingness to continue learning and participation in the excitement of progress. At this stage few young people who do not proceed to higher education are sure of their ideas for the future, and most have had no experience of work in engineering. Also, no easy measure of attitude is available. Therefore, whilst personnel staff can perform an initial selection based upon recorded aptitudes, the final selection must be made by engineering managers who understand the needs of the business and can discern the attitudes of the people they interview.

The attitudes that indicate whether young people are likely to succeed in engineering are reflected in their personalities and in their interests. The interviewer should look for confidence and evidence of an outgoing personality. Young people starting in engineering direct from school who lack confidence and verve are seldom successful, since at the levels of work they will undertake they will almost certainly be junior members of teams, and they will be expected to

---

* Sadly, this route has almost disappeared in the UK, though the government-sponsored "modern apprenticeship" scheme is an attempt to recover from the damage that has resulted.

contribute to the team effort. They are also likely to be given a series of fairly short assignments, so they must be able to adapt quickly to new people and new problems, learning as they go.

Graduates

The selection of graduates into engineering must be based on the same general principles of assessing aptitudes and attitudes. Now, however, we have evidence of both from the college or university results. Currently many engineering graduate courses include periods of work in industry. However, most graduate engineering curricula are limited in their coverage of practical topics, such as actual machine tool use or technical details of electronic component construction and test. The attitudes of graduates to engineering work can be tested by asking what technical journals or magazines they read, what engineering society activities they have participated in, and what practical work they have done, for example car maintenance or hobbies. A graduate with a good engineering degree but who cannot provide any evidence of interest beyond the curriculum should not be preferred to one whose academic qualifications are lower, but who is better equipped in other mental departments.

Managers can improve their chances of selecting the best if they work closely with selected universities and colleges and maintain contacts with their staff and with students during their courses. Then attitudes can be tested early and encouragement can be given by visits, lectures and interviews with selected students. University and college staff will contribute to this effort if the company in turn supports them, for example by providing external lecturers or equipment, scholarships or research funds. This kind of expenditure might appear to be largely altruistic, and therefore subject to close scrutiny when costs are being reviewed (or, even more problematical, when being initially proposed), but whilst the long term benefits can never be quantified, they can be enormous.

It is often necessary to recruit non-engineers into engineering teams. Mostly these are graduates in subjects such as physics, mathematics, chemistry, metallurgy and computer science. The same care must be applied to their selection, though they are likely to be employed in more specialist roles. Nevertheless, it is important that they can take an interest in the wider problems, so that they can appreciate their contributions and understand the effects of their work on projects. A brilliant problem-solving specialist might be of great value, but if the nature of the problems change to areas outside his specialization he must be willing and able to adapt and to continue to learn. Non-engineers working in engineering teams should always be considered for further training in engineering subjects: for example, mathematicians and software scientists working in system development can benefit by short courses in basic electronics, which will enhance their contributions as well as equip them better for future, more senior appointments.

Experienced People

Experienced engineers are equipped with the records of their qualifications, as well as of their experience. Their CVs should tell the whole story of their academic and work experience and results. These might be supported by references and other information. Interviews should therefore be concentrated on the less tangible aspects of personality, ambition, integrity and interests. Experienced people sometimes have constraints related to location, situation, health and age, all of which must be taken into account. The interviewer's experience and intuition must be applied to balancing these factors. Generally speaking, the applicant's attitude and motivation in relation to the job and any related problems are more significant than the problems themselves. This applies particularly in relation to age: it is important to remember that a person of, say, 50, has probably 10 to 15 years of productive capacity ahead, which must be balanced against the possibility that a younger applicant might leave earlier. The additional experience of the older applicant could be extremely valuable, and older people often show surprisingly youthful enthusiasm coupled with maturity and loyalty.

Selection From Within

As far as possible, selection for positions other than for the lowest levels should be made from people already within the organization. Vacancies should always be advertised internally. Every vacancy at levels above the lowest should represent a chance for promotion, and the people in the organization should be given every opportunity to improve their positions. Conversely, every placement from outside is an opportunity lost, and the people who consider that they were qualified for the job will resent the appointment and will feel betrayed. There will be occasions when no internal applicants will be suitable, and managers must be free to judge this. However, the policy of internal preference must be clear, both to the managers concerned and to all other employees.

Employees at all levels must be encouraged to qualify themselves for internal selection and promotion, and managers must always work to improve the qualifications, experience and prospects of their people. There is a strong tendency among many engineering managers to hold on to their staff, fearing the problems that would follow if they left. In furtherance of this, they are reluctant to approve training, since they consider that the improved qualifications will encourage people to leave for better prospects elsewhere. This negative attitude to training and development is bad for the organization as a whole. It is also counter-productive for the managers involved, since their staff turnover is likely to be higher, not lower, as their people realise that their only chance of advancement is outside the organization. In addition, the average quality of the people who remain will fall, and motivation and therefore performance will be low.

If every manager encourages training, development and advancement there will be a continuous flow of people across and up the organization, all gaining in qualifications and experience, and all functions and projects will benefit. Morale, motivation, and loyalty will be improved, and the whole organization will respond as a team. The institution of a firm policy for internal selection is one of the most effective methods for improving long-term performance, and it costs nothing.

Managers

Engineering managers must never be selected young. Young people (school leavers and graduates) have not had the opportunity to demonstrate the talents necessary for management and there may be nothing in their records to indicate that such qualities are latent. (There might be some early indications, such as leadership roles in sports teams, clubs, etc.). Measurable qualities, such as IQ, examination results and various psychological tests can provide only very tentative indications, and then only for a limited promotion horizon. Despite this, many organizations make selections, mainly of graduates, as "management trainees". They are told that they are destined for senior management positions, and this is also communicated to those less favoured. The anointed ones are given special treatment, including experience of different functions and further training, and their promotion is guaranteed up to a certain level.

This is a sad and destructive policy. It inflates the expectations of those selected and causes resentment among those who are not. It is particularly damaging in organizations that run internal training programmes, such as apprenticeships for school leavers, some of whom will show management potential that exceeds that of the selected graduates. Experienced people further up the line also feel resentment at the special treatment given to the management trainees, particularly when the trainees displace them in selection for jobs, or become their bosses.

It might be argued that graduates with good degrees would not wish to join organizations that do not offer enhanced promotion prospects. In fact, the most suitable graduates for future management positions seek opportunities, not guarantees. They will not reject offers of employment that put them in competition with others in the organization. Other aspects of the offer, particularly the overall performance and reputation of the organization as an employer, should be more important and should be stressed. It is also important to have in place opportunities for advancement that are not necessarily related to management, as we will discuss later.

Some organizations go to great lengths to identify management potential. The armed forces' officer selection procedures subject applicants to several days of personality tests, exercises and interviews, followed by rigorous training for several months or years. Despite this, the forces do not guarantee promotion to those who succeed, beyond quite low levels. Organizations that attempt to select future managers on the basis of an interview and examination results, sometimes supplemented by a few tests, might as well rely on horoscopes.

Another destructive policy is the not infrequent practice of senior managers, usually at director level, to appoint young engineers as "personal staff". In the course of their work they are given access to information that might be denied to lower level managers. They often attend high-level meetings and, as the messengers of the senior managers, they are accorded a status incompatible with their experience and seniority. For the individuals selected the roles can be interesting and illuminating, but their position is invidious.

If a senior manager does need this kind of support it is much better to give the opportunity to a number of people, by making the assignments short term ones. In this way the job can be seen as part of the development process for young engineers, and not as a privileged position for a very few. More engineers will be given an insight into higher management and there will be more trust and cooperation.

Such appointments are, however, best avoided altogether. They cut across management responsibilities, so if a task in support of the senior manager is necessary, it should be assigned to the appropriate manager or to a management team. Young engineers should be given experience of management judgments and decisions, but it is better for this experience to be gained incrementally and at levels not far above their current responsibilities.

The only effective and fair way to select managers is to give all employees the same opportunities to demonstrate management talents, one step at a time. These talents must be identified by performance in the present job and developed by training. As management vacancies occur, all qualified staff must be encouraged to apply. Qualifications for each position must be stated, so that everyone knows their chances for selection. In this way the organization will ensure that all of the most suitable people are available for each promotion, and that those promoted know the procedures, methods, and people in the organization. It will also ensure that promotions are seen by all of their people as being fair and open, so improving morale, loyalty and motivation for self improvement.

Higher Management

The principles described above apply also to higher management. Senior managers should, whenever possible, be selected internally. If suitable people are not available for promotion the system for management development should be reviewed. There might be special reasons for taking on a manager from outside, particularly when new methods are being used for the first time, but it is important that the appointment is welcomed by colleagues and by the people in the new manager's team.

Sometimes a senior appointment from outside can bring in fresh ideas, particularly if the appointee has a good reputation in the field. He will have no loyalties to others below him, so he might be more objective in solving staff problems. These justifications should not apply, however, in organizations that develop the management potential of all of their people and manage them according to the principles described earlier.

TRAINING AND DEVELOPMENT

All engineers must be given regular training to ensure that they are up to date in their specializations, that they widen their knowledge, and that they are qualified to manage engineering work. The extent of training for each objective must be related to the individual's objectives and capabilities. For example, an electronics engineer might be excellent at circuit invention and design and might not want to move into management roles. His training should then be concentrated on ensuring that he remains up to date in his field, and also that his knowledge is broadened so that he can work better across interfaces and is available for related work if necessary.

Training can be presented in many ways. Formal courses, from full-length postgraduate courses to short specialist courses, can be used. Engineers should also be encouraged to learn on their own, by experience, by reading books and articles, and by attending seminars and lectures. Managers should also organize internal training, to pass on their own knowledge and that of specialists within their teams. Training must cover the two prime topics: engineering knowledge and management. Management training can include other topics, such as languages useful to the organization.

The great majority of Western companies provide insufficient training for their people. The training budget is classed as an overhead burden, and time away from the job is considered to be mainly a diversion, and always an inconvenience. These are, of course, short-term considerations. Training is not a burden, but an investment. The short-term local inconvenience should be rewarded by more productive work later, across the whole organization. Training policy must therefore be made a long-term commitment, and training must be given priority on par with all but the most urgent short term tasks. Both Drucker and Deming emphasized the importance of continual training for all staff. As a result, most Japanese engineering companies devote considerably more resources and time to training than do their Western competitors.

Records must be kept of all training, whether long or short term, formal or informal. The individuals should be given copies of the training records. These, combined with records of experience, are essential for planning and selection.

To avoid the problems of staff leaving for better-paid jobs as a result of receiving training, and also to recognize its value, pay increments should be linked to training received. One way to achieve this is to allocate points to approved engineering and management courses and qualifications, on the basis of value to the organization and the amount of personal effort involved, and then to link annual increases to points achieved.   Furthermore, qualifications for promotions could be linked to completion of defined training, as is normal in the armed forces, thus ensuring that those promoted have the knowledge necessary and at the same time encouraging staff to improve.

Development must include experience in different parts of the organization. There should be a definite but flexible policy on internal transfers to ensure that engineers are not retained in one job or function for more than a few years. The time should depend upon the type and level of the job. For example, a graduate should be moved every 6 months or so, but a senior specialist or manager could hold a position for 2 to 4 years. Transfers should include time in as wide a range of engineering functions as possible. This ensures that people develop experience of the whole organization and that managers have the breadth of knowledge essential for their work. Again, the policy of providing work experience is most highly developed in Japanese companies. For example, in the car industry it is normal for engineers to have to spend up to 6 months actually working on the production line before starting work in design.

SENIORITY, PROMOTION AND SALARIES

Seniority should be related to age and experience. Companies in retailing, finance, and other non- engineering sectors can be managed by people with lots of flair but little experience, since these industries are not based upon changing knowledge and multidisciplinary teamwork to the extent that engineering is. We do observe excellent young senior managers and chief executives of engineering companies, but these companies are usually specialists in new technologies or applications, in which the only experience resides in these young entrepreneurs.

Sufficient experience to manage engineering in a mature industry cannot be developed over a short time. Also, people in teams prefer leaders with experience. Very few young people have the talent and maturity necessary to lead a team of experienced older people effectively, particularly if they have not been carefully selected and trained.

Promotion must not be based entirely on the principle that seniority and status come only with management roles. This idea is prevalent in much engineering industry, so that good engineers frequently end up as mediocre or bad managers, or they remain unpromoted and therefore dissatisfied. The value of engineering skill and knowledge must be recognized by the existence of a separate route to high rewards for professionals who are not managers. The very best and most experienced can be given consultancy status, at a level equal to senior managers. Senior professionals must be expected to contribute to projects, maintain and develop their knowledge and provide training. They should be recognized in the organization and in the appropriate professional community. Promotion must be open for both engineering and management roles, with flexibility to allow people to move between the two. This means that rigid categories, for example "career grades" and "job grades", should be avoided. People should be considered for jobs on their own merits and on the requirements of the job. Ideally, most engineers should have experience of both types of job, and of course most engineering jobs combine both aspects in varying degrees.

Salaries should be based on the job, and on experience and qualifications. Since experience is developed over time, time in the organization must be rewarded. Salary increments should be awarded to all who perform their tasks satisfactorily, every year or two, in addition to general increases to cover inflation or profit sharing. Not only does this recognize the value of experience, it also recognizes that people's commitments usually increase with age, and it therefore reduces the pressure on good people to seek more highly paid employment elsewhere.

Deming explained why the practice of "merit increases" is counterproductive and demotivating. Such awards are based largely on subjective judgments of the performance of people working within the system and as members of teams. Individual performance can be greatly influenced by the management system, for example in terms of resources supplied, training given and leadership. Also, it is extremely difficult, and often invidious, to separate individual from team performance. Those who receive above average increases will be pleased. Those who receive below average or no increases must be told that their performance did not merit more, and this is invariably highly demotivating. Very few people are prepared to accept that their performance is unsatisfactory. When the whole team performs well the manager should never be put in the position of having to tell some of them that their performance will not be rewarded.

Salary progression should therefore be based only on experience, knowledge gained, and the responsibilities of the job. Increases should be paid when engineers are transferred to other jobs as part of their development and when the transfer is to a higher position, either in terms of engineering or management responsibility.

Additional performance-related payments can be made for specific reasons, such as a project team achieving a major objective or the company exceeding a profit target. These should be one-off payments, not salary increases. It is always preferable to reward all in the organization, rather than a particular group, to encourage the feeling that everyone is contributing. If a project team achieves an objective for which the whole organization will be rewarded the team should receive some extra non-monetary recognition, such as a celebration party or letters of appreciation.

OBJECTIVES AND APPRAISALS

Deming argued against the common practice of setting measurable objectives for the performance of individuals, and against having formal, regular appraisals for monitoring performance against the objectives. This advice is considered by many managers to be controversial, and few organizations understand it or apply it. Objective setting ("management by objectives" (MBO)) and formal appraisals are the norm in most engineering companies. It seems obvious and rational that people should be assigned performance objectives and then be measured against them, because that is what we do with machines and processes. In other words, that scientific principles should be applied to the performance of people. Drucker actually teaches that objectives should be set and people appraised. Thus the two most influential teachers of management practice appear to disagree. How can we reconcile these views in the context of engineering?

Deming argues that people will tend to concentrate on achieving those objectives against which they will be appraised, regardless of the fact that the criteria to be demonstrated might often be incompatible with the best interests of the business, and will often be open to different interpretations. People will be reluctant to use their initiative or to perform other tasks if these will jeopardize achievement of the objectives against which they will be appraised. These problems are particularly likely when the objectives are expressed quantitatively, for example "reduce production defects by 20%". A manager faced with such an objective might work to achieve it, but in the process might neglect equally important other work on planning the evaluation of a new process. He could also achieve the objective by redefining the defect criteria, making no real improvement in quality. Almost any specific objective, particularly those that are quantified, can be subverted in such ways. Other examples that occur are "complete 90% of tasks on time" (Are they completed well, or just completed? Were the task schedules easy or difficult? Did the priorities differ?), or "maximize the proportion of time booked to projects" (is the time spent well, or just spent?), and even "publish more research papers" (regardless of quality or originality?). This last seems to apply strongly in academia!

Note how the objective "maximize the proportion of time booked to projects" can lead to inefficiency. It discourages people from completing work more quickly, if the time that is then made available has to be booked to other useful but non-project activities such as learning. Quantitative objectives, "targets" or "key performance indicators" (KPIs) such as these are frequently applied, and they are usually counterproductive.* They provide misleading criteria for relevant performance and they discourage both managers and managed from thinking and doing what is best in the circumstances.

Sometimes identical objectives are set for a group of people doing similar work, for example managers of functions, who are then judged on the results. This can focus their concentration on achieving the objectives, generating competition between them rather than cooperation.

Deming pointed out how the variation inherent in a process, and possibly out of the individual's control, can make it appear that an objective has been attained or missed without any action on his part. A numerical objective is therefore meaningless if the process is not under the individual's control, and if it is not accompanied by a specific plan for achievement.

In effect, Deming argues that the ideal objective for an individual manager or engineer is "do your job well, and always better, to further the aims of the organization". Then everyone has the same objective, which is totally in line with that of the team and of the organization. Adopting the sports analogy, a team manager does not specify how many points or goals individuals must score, as this would discourage teamwork. Instead he motivates and trains them to win games. When the objective is as general and as all-encompassing as this it becomes unnecessary to state it formally, and specific, individual objectives become superfluous and counter-productive.

---

* Governments have become enthusiasts, setting targets for health, education, crime and many other aspects of public life. Have you noticed the improvements?

Despite these strong arguments against the general principle of management by objectives, specific objectives can sometimes be appropriate and can help to focus attention on particular needs for action or improvement. The prime requirement is that all objectives must be closely allied with the aims of the enterprise. They must also be clear and unambiguous, so that they cannot be misunderstood or subverted. If they are subject to variation, the sources of variation must be identified and allowed for. Therefore objectives must be very carefully worked out and accompanied by an agreed plan for achievement and if necessary measurement, to ensure that the objectives are achievable, the methods to be used are understood and accepted, and that the processes involved are sufficiently under control. They should be considered as informal contracts between the individuals and their managers, with both having responsibilities to one another for their achievement.

The objectives must never be exclusive of other work that is to the benefit of the enterprise and within the competence and power of the individual. They must never discourage initiative, but rather must positively encourage it. They must be flexible, to enable changes to be made if necessary. Flexibility must extend to timing: objectives should not be set only at convenient annual intervals, since engineering and business opportunities and problems are not constrained to follow the Earth's rotation around the Sun. These requirements militate against the setting of general objectives for groups of people at set times.

Therefore appropriate objectives could be "generate significant improvement in product quality, while continuing to support design teams", "reduce the time to generate detail designs that are correct first time", or "complete a course on fracture mechanics and report on its value to the project". Objectives expressed in terms such as these, and the plans to achieve them, can be discussed when they are set, and later, and performance against them can be judged intelligently in terms of their value to the organization and to the individual, rather than in isolation.

Individual and team objectives must be defined separately, though of course the individual's objectives must be in harmony with those of the team. Therefore individual objectives must be related to tasks that are allocated to people, rather than to the team, and to personal development needs, such as success in training programmes. When the individual manages a team, then the team objectives coincide with the individual's task objectives.

Since engineers nearly always work as members of multi-disciplinary teams, dealing with problems that change as projects proceed and technologies develop, individual objectives must relate only to the contribution to the team and to the needs for personal development. If a test engineer is told that, as a prime objective, he must have a test system designed and installed by a certain time, he could achieve this but it might be unreliable and cause delays and additional costs later. His objective must be to have the test equipment ready when needed, but all other relevant aspects of the task must also be taken into account. It might be better to delay the installation to enable the design to be reviewed more closely, or to allow an aspect of the specification to be clarified. Some aspect of the project outside the engineer's control, such as delivery of an instrument, might cause delay. It might even be decided to speed up provision of the test installation by switching resources from elsewhere. The engineer must be involved in the team activities of planning, implementing, reviewing, and reacting. Any objectives that he is set must reinforce all of these aspects of the job and must not detract from any. Since all the activities of the team members interact with one another, in ways that are difficult to forecast, trying to set specific objectives for each member can be difficult and can lead to a breakdown in teamwork. The comments made in Chapter 6 on project planning using PERT relate also to individual objective-setting in engineering.

Objectives for personal development can, of course, conflict with work objectives in the short term. Therefore, like any other objectives, they should be given the appropriate priority and a plan must be agreed for achieving them. Once agreed they must be actioned: training courses should be reserved and time made available, so that job responsibilities can be planned around them. The manager normally carries the greater responsibility for action on development of his people than they do themselves, because only the manager can set the priorities and make the time available for training in working time. When the individual is assigned to a project from a functional group, it will be necessary for the managers concerned to work together to agree on the priorities and timings.

Appraisals of the performance of individuals should be a constructive exercise, to review progress against objectives and to help to plan future work and development. The appraisal of progress must, of course, be related to objectives that are set according to the principles discussed above. The appraisal must be a two-way discussion to determine the extent to which the objectives have been achieved and the reasons for any under-achievement. Actions necessary for improvement must be identified, agreed and planned in respect of the individual's performance and of any other relevant factors, such as problems in the system. The individual's overall contribution to the team should be discussed. The appraisal must also identify needs for personal development, both for the short and long term, related to the objectives of the team and of the business. Professional engineering aspects of performance and development must be dealt with separately from management aspects, again in relation to the requirements of the job and of the individual's talents and aspirations. Many engineers are excellent professionals but less adapted to or enthusiastic about management. Many others are less skilled professionally but are natural leaders. These varying talents must be identified and appropriate plans must be made for development, recognizing and exploiting the strengths identified without criticizing the other characteristics.

Appraisals must never be admonitory. If there is a need for admonition this should be handled completely separately, at the necessary time, and not saved up for the appraisal. Any admonition during the appraisal will destroy the cooperation necessary to review progress and to point the way to further objectives and improvements. On the contrary, every opportunity should be taken to recognise effort and give praise. The individual being appraised should be encouraged to comment on management aspects that influence his work and development.

Appraisals must not be used as the basis for salary adjustments and "merit rises". As explained earlier, salaries should be based on responsibility and experience, and the practice of awarding increases to selected people on the basis of individual appraisals is nearly always highly damaging to morale and teamwork. No manager should be put into the position of telling an engineer that he has performed well, but that he does not qualify for this year's merit quota. Like an admonition, this can destroy the relationship necessary for a successful, fruitful appraisal interview.

Appraisals must not be used as a criterion for dismissal. There have been well-publicized stories of companies systematically firing an arbitrary number of the lowest-scoring people. This is a terror tactic, absolutely contrary to the new management philosophy. Sometimes it is necessary to dispense with people, for poor performance, disciplinary reasons or redundancy, but these situations must be dealt with when they arise and on their merits.

A great variety of paper forms are used for recording appraisals. These range from very simple to complex, multi-page documents. In many companies the forms change from one extreme to another, depending upon the ascendancy of different points of view on the appraisal process or the whims of personnel or "human resources" managers. Most engineers like simple forms, but many personnel professionals and consultants advise the use of more complex systems. Any tendency to complicate the record, particularly if it tries to quantify performance and attributes, runs the risk of becoming too "scientific". The system can end up serving itself and not the people or the organization. It is essential that the system used is accepted by the people who will use it, and not imposed upon them without discussion and training.

Some appraisal record systems require indices to be used against the attributes listed, and others concentrate on verbal inputs. As discussed in Chapter 6, people are not amenable to easy classifications as are, for example, the cutting speed of a machine tool or the frequency range of an oscilloscope. Human performance is highly variable, depending on attributes, leadership, training, experience, mood, health and other factors, and to varying degrees these are influenced by the system in which they work, their managers, their peers and others, including their families. Almost any judgment of performance is also subjective and conditioned by the appraiser's perspectives, attitudes, etc.

With such scope for uncertainty and subjectivity it is not surprising that appraisal systems are so diverse and controversial. However, there are some basic principles that can be followed to minimize subjectivity and to ensure that the potential benefits of systematic appraisals are achieved.

Appraisals can give managers and their team members valuable opportunities to add value to one another and thus to the whole organization. They can foster growth in the capabilities of people, in professional and in wider management skills, and they can help to build trust and honesty between the people involved. However, appraising other people in a face-to-face meeting is a task that many managers find difficult. It is an intensely personal encounter, requiring honesty, firmness and some courage. It is not the sort of activity that comes naturally to most people, so the skills needed for effective appraisal must be developed.

Effective appraisal must be based on modem principles of managing people, as discussed in Chapter 3. In particular, "scientific" ideas must not be allowed to influence the dialogue. Appraisals must also be consistent within an organization and there must be stability of the system used. Consistency can be provided only by having a stable system, supported by adequate training. Organizations that maintain this combination, particularly the armed forces, have appraisal systems and records that work effectively because everyone understands them and therefore they are generally trusted. It is much more important to have a stable system, used by people who have been properly trained in its use, than to seek repeatedly to develop "better" systems.

Deming's arguments against formal appraisal systems are based on the same reservations he states against setting objectives. Ideally, managers and their people should be in continuous dialogue and harmony, and an excellent manager should not require the support of formal appraisals at set times in order to direct and develop his team. Conversely, those that are less proficient might do more harm than good by performing appraisals that do not follow the principles described. However, in large and medium-sized organizations it is necessary to maintain a record of people's work and development so that they can be best used and improved. Even in small organizations such a record should be valuable, since people's memories cannot be relied on. Since engineering involves a wide variety of skills, experience and tasks, and since continual training and development is so important, a systematic approach to recording information on the experience and development of engineers is nearly always advisable to assist with selection for new jobs and for promotion. The system should contain as a minimum the record of qualifications, all assignments and all training.

In conclusion, Deming's criticisms of objective-setting and appraisal are a logical extension of Drucker's "new management", in which the ideas of "scientific" management are exposed as being inappropriate for managing people. Objectives and appraisals can be, and often are, counterproductive and divisive. Nevertheless, if they are carefully planned, aligned with the needs of the business and emphasize individual development, they can be used to provide a foundation for improving efficiency and productivity. To minimize the possibilities of misuse they should be kept simple and stable. Most important, managers must be carefully trained in the principles and the human skills of objective-setting and appraisal, to ensure that consistency is maintained and that the system works to the benefit of the organization and its people.

THE PERSONNEL MANAGEMENT FUNCTION

It is unfortunate that personnel (or "human resources" (HR)) management, as a specialist function, has developed into a form of meta-management, based largely on distorted ideas of "scientific" management. Drucker specifically warned against this development, yet it has taken root in many organizations and it is proselytized in books and by consultants.

The most extreme example of this trend is "psychometric testing". There is a wide variety of such tests, ranging from interpreting inkblots to answering questions related to personality, and IQ tests. A typical, published claim is that "the likelihood of identifying an above-average employment candidate by random selection is 15%. When psychometric testing is applied the probability increases to 76%". Apart from the obvious question of exactly what criteria determine an "above average candidate" (simple statistical theory dictates that 50% of any random sample must be above average), the argument is specious and misleading on a more fundamental plane.

No manager would select staff for jobs or promotion "randomly" or blindly. For every selection a large amount of information on the candidates already exists, as described earlier. Aptitudes, attitudes, experience, and knowledge are all described. Even more important, the candidates are known as people. The knowledge might be based on years of working together, or on a brief interview. Whatever the experience, aspects of personality that cannot be measured can be observed. Selection is not a one-way process: the candidate will respond differently to different situations, such as the nature of the job being considered, the reputation of the organization as an employer, recent employment experience, and who is to be his new boss. Psychometric testing is a one-way process, a measurement of particular attributes in isolation from the actual world of work and from the other people involved. It is like measuring the accuracy of an instrument in a temperature-controlled, vibration isolated laboratory, then expecting it to be as accurate when used by field engineers working in harsh outdoor conditions.

Personality and attitudes are formed and developed by experience and by relationships with others. This process is not frozen at the time of the psychometric test. Personal qualities are not static, like the performance of a machine. We all know of people who perform badly in a particular role, or whose managers say this, but who later do a very good job elsewhere. The reason for the difference is often the complex interplay of personalities between the manager, the individual and the rest of the team. Ultimately, the manager must carry responsibility for obtaining the maximum performance from the whole of his team. Psychometric "data" cannot help him to do this. It might if he were playing a game, in which the "people" are tokens, but it is a cowardly approach when real lives are involved.

Other examples of over-enthusiasm for the specialist fads of personnel management are the development of complex appraisal systems and the imposition of training requirements that are not supported by the productive functions of the business. In extreme cases, external consultants are brought in to conduct tests and appraisals on people already within the organization.

The personnel management function can provide very effective support to the aims of the organization if it restricts itself to administration and avoids policy intervention. Policy on people must be made by managers, starting from the chief executive. By acting in support, for example by organizing and administering training, recruitment advertising, dealing with formal and statutory procedures, etc., it can relieve managers of much detail and help them to concentrate on managing their people and their tasks.

The personnel management function need not be passive, however. It should take an active role in recommending training, in helping to improve management and professional skills and in assisting with policy-making. It can be engaged in the selection of young people who will be given work in different parts of the organization as their careers develop, particularly in large organizations. In this particular case, when selection is being made for a variety of future roles, psychometric tests can be useful supplements to other information, but actual performance should provide the criteria for future development.

For these reasons it is important for the personnel manager to be familiar with the business, and not just a "personnel professional". In an engineering business this task can be very well performed by an engineer with a good management and training record. The personnel manager's position must never be elevated beyond that of the senior managers he supports. Titles such as "human resources director" place the role at too high a level, whilst implying a "scientific" approach to managing a "resource". Every person in the organization is unique, variable, and possesses great potential. Only their managers can exploit and develop these capabilities.

CONCLUSIONS

The development of engineers starts at primary school and continues for life. Engineering is not unique in this respect: good accountants and doctors must also have the right initial skills and must continue to improve. However, engineering is subject to more rapid change and depends more on multidisciplinary teamwork than any other profession.   Therefore the importance of continual development is enhanced, and attitudes and leadership skills are more critical to success.

Managing the people in the enterprise is the most important role of management, at all levels. It is also the most difficult, and the least supported by science and mathematics. Performing an investment appraisal for a new machine or running production options through a spreadsheet are simple, mechanistic tasks, with predictable inputs and outputs. The effects on short term and long term performance of training programmes, approaches to appraisals, remuneration policy or informal talks with individual engineers are not amenable to such analysis. Yet engineers are analytical people: we have had scientific and mathematical educations. We enjoy analysis, and the results give satisfaction and a sense of achievement. They are incontestable, or at least provide a strong basis for argument. Engineers are therefore prone to be frustrated by situations that are not amenable to quantitative analysis, such as the behaviour and performance of the people working for them.

This is the greatest challenge for the managers of engineers, particularly as most of them must be engineers themselves. Concentration on managing the system and then expecting performance as a result will not release the potential for improvement that resides in the individuals and the teams. Note how different this is from managing a robotic production line: the robots have no capability for self-improvement, either individually or together. We must, as the first priority, manage the people and the results will then flow. In order for the development of the people to be at all effective there must be trust and confidence in their competence to perform their tasks and make decisions. Management development must include the generation of trust between managers and managed, and across the organization. Expectations of performance can then be based on the confidence in the people concerned, rather than on pseudo-analytical forecasts that take no account of the great versatility and variation of people at work.

All decisions related to the development of engineers and engineering managers take time to become effective, and their effects extend for many years into the future. As explained above, we cannot quantify these effects, either as forecasts or retrospectively, since the cause and effect relationships are not deterministic. However, the decisions on selection, training and promotion are the most important for the future of the enterprise. As Drucker wrote, the only aspect of the future that we can forecast with confidence is the need for good managers, whatever the conditions.

5

ORGANIZING ENGINEERING

CULTURAL INFLUENCES

The way that engineering work is organized, whether it is revolutionary or evolutionary or shades of grey between, is strongly influenced by the culture and training of the people involved. The cultural heritage of Western civilization, in terms of the way we think about problems and develop solutions, was formed by the early Greek philosophers such as Plato and Socrates. From them came the ideas that truth is found by logical argument, if necessary by demonstrating the illogic of opposing ideas. They also developed the ideas of reductionism: problems and methods should be reduced to the smallest possible understandable elements. If these can be understood and controlled then so can the whole. Western society's approach to argument, debate and organization has been formed by this tradition. We take for granted the methods that have been passed down by earlier generations, and we find it difficult to accept ideas that seem to run counter to the logical, structured view of the ideal world.

All engineers have been educated and trained to think rationally and scientifically. Since all engineering is based upon the principles of science and mathematics, it seems rational to apply these principles to the management of engineering. Engineers instinctively feel uncomfortable with systems that appear to be disorganized or not under tight and visible control. It does not follow that all engineering is subjected to such controls: apparently haphazard systems are not uncommon, especially in small, new businesses. However, they are disparaged in comparison with others that have comprehensive and visible systems of organization and procedures. Obviously any productive enterprise must be based on some system and must have controls, or there will be no objective and no way of making progress towards it. A totally haphazard group of people will not be able to develop an engineering product, far less to do so better, more quickly and more cheaply than better organized competition.

Where then lies the optimum balance between system and chaos? The traditional answer has been that this is towards the maximum of order and control. Drucker, Deming and others have taught that looser, less formal structures can be more productive, and companies that have followed this teaching have excelled, as described for example by Peters and Waterman in their book *In Search of Excellence* [7]. However, many companies and other organizations remain strongly under the influence of "scientific" management, and this is particularly the case in engineering. Yet it is interesting and ironic how often we observe bad management of engineering companies and functions, manifested in bureaucratic but shifting organizations and procedures, low morale and poor productivity. When crises occur the standard reaction is to change the organization and the procedures in the direction of greater accountability and control. Seldom do such situations lead to looser, less formal organizations and greater freedom of action for managers and individuals.

TRADITIONAL MANAGEMENT

The traditional approach to management of enterprises is based on our perceptions of rationality and scientific reductionism. Thus we divide work into specializations, starting with our approach to education and training. In engineering, we create functional organizations based upon these specializations, such as mechanical, electronic and software design. Each function is managed by a specialist in the field and separate mini-cultures develop, manifested by salary differences, working facilities and so on. Within the functional specializations sub-specializations develop, such as structure design or digital electronic design.

Specialisation also extends to the timing of the different kinds of work involved in product evolution. Conceptual or system design, detail design, test, production and support are considered to be separate functions, and the results of the work of each are passed to the next in the evolutionary line. Since the early phases are seen to be the most forward-looking and innovative, they attract higher status and esteem. Conversely, people at the production and support end are supposed to be doers rather than thinkers, working to the designs passed down to them. They are not expected to have the same levels of intellect or innovative skill, and so attract lower levels of recognition. Recruitment to the different functions follows the same pattern, with highly qualified new staff being assigned to the 'leading edge" functions. Since the managers of the functions usually rise within them, the imbalance of capabilities is maintained to these levels, so that the managers of the downstream functions are often considered to be of lower status than those of, say, the system design function. The inevitable consequence is that the downstream activities are not performed as well as they could be. Also, since the conceptual, system and detail designers are not closely in touch with the seemingly more mundane problems of production and support, their designs often do not optimise the product for these aspects, leading to production and maintenance problems, delays, higher costs and lower competitiveness.

Higher managers have the difficult task of coordinating the work of the specialist functions, in terms of their contributions at all times during the evolution of the product and the correct sequencing of their work as the project proceeds through the phases of design, development, manufacture and use. They must apply the principles of planning, organization, delegation and control, as taught by conventional training and as interpreted by higher levels of management. All levels of management want "visibility" of these aspects, including the ability to measure performance against targets.

Plans are necessary to determine how objectives are to be achieved, what resources will be required, the timing of activities and responsibilities for them. The more detailed and precise a plan, the less scope there is for differing interpretations or uncertainty. Plans are then subdivided into lower level plans for subordinate activities. A good plan is considered to be one that leaves no scope for interpretation or uncertainty. In engineering, plans include specifications.

Some sort of organization is necessary to determine responsibilities for functions. Organization charts are used to indicate functional responsibilities and levels and spans of management. All people in the organization know their responsibilities and those of others. They also know what they are not responsible for.

ORGANIZATIONAL FORMS

Three basic forms of organizational structure are found in most companies. These are functional, project based, and matrix, as shown in Figures 5.1, 5.2 and 5.3, respectively.

The functional form of organization places each engineering function, such as system design, specialist design, manufacture, product support, etc., as well as support functions such as accounts and personnel, under separate managers, who are then responsible for supporting projects by assigning people to project tasks. The functions are responsible for the quality of this support, and must therefore recruit, train and control the staff so that they can provide the best support to projects. However, when several projects are being supported there can be problems of coordination and communication. It is also difficult to generate a team approach to projects, and loyalties tend to be split between the functions and the projects.

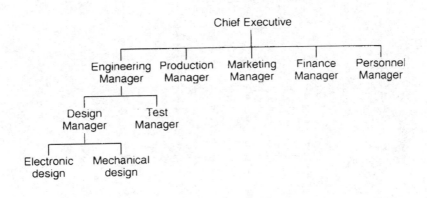

Figure 5.1 Functional organization

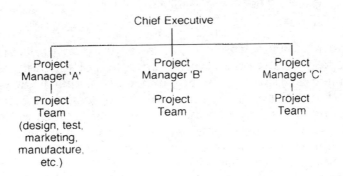

Figure 5.2 Project based organization

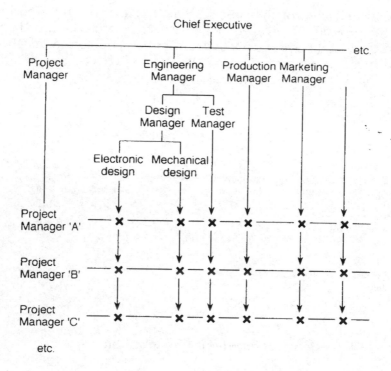

Figure 5.3 Matrix organization

The opposite approach is to organize entirely on project lines, so that all people working on a project, regardless of function, report to the same manager. This has the advantage of making teamwork and communication easier and it is easier to control work, since functional boundaries can be largely eliminated. However, other problems are introduced, such as balancing work between different projects, assuring common approaches to specialist tasks when appropriate, and development of specialist staff.

To obtain a balance between the advantages and disadvantages of both types of organization, most companies adopt compromises between them. Matrix management is a form of organization that is basically functional, but with an overlapping project structure in which project managers draw support from the functions by having functional staff assigned to them as required. In this way the project manager is given responsibility for the success of the project, and functional   managers   retain   responsibility   for   recruitment, development and techniques, including the issue of procedures relating to their functions. The project manager is usually further supported by administrative staff, such as project control and secretarial. The main problem with matrix management is the fact that the people working on projects have a dual loyalty, to the project and to the function. It is also necessary for project and functional managers to work in close harmony, and this can be difficult when project and functional priorities are in conflict or when different projects are being supported, all demanding priority.*

Delegation of tasks follows from the organizational structure. Each level of management ensures that the tasks necessary to perform the overall function are allocated to teams or individuals with the appropriate skills and other qualities, and this process is continued downwards until all levels of the organization are involved. Terms of reference for departments, teams and individuals further describe, and circumscribe, what is expected of people.

The whole process must be controlled, and control involves a variety of techniques. The methods used by the various contributing functions must be harmonized where necessary. For example, different design teams on a project should use the same criteria for selecting components and processes, and they should all take account of the assembly and test methods that might be used in production. Therefore they should work to common standards, which would usually be a combination of external (national and international standards) and internal procedures. Such standards and procedures also help to determine the quality of the work involved, since they often (but not always: see Chapter 10) represent best practices based on experience. They provide templates against which the quality of work can be judged and they provide guidelines for the less experienced.

Controls are also necessary to ensure that all activities are performed in the correct timescales and sequences and within allocated budgets. Project management methods such as Programme Evaluation and Review Technique (PERT) (also called critical path management (CPM)) and PERT/COST have been developed and are available in computer software, both as individual programs and as parts of larger integrated business systems such as SAP™ and Oracle™, so that changing events and relationships can be quickly

---

* Of course, if there is only one project the organization would be based around that, and most of the options would merge.

input and analyzed in relation to their effects on project timescales and costs.

All of the methods of planning, organizing, delegating and controlling can be taught and explained. Books and management courses describe the principles and interpretations of them, and individual managers apply them within companies and other organizations. Such structured systems and procedures can be demonstrated to others, such as higher managers or those performing assessments of management systems in accordance with the practice of quality system assessment (as discussed in Chapter 10). An organization or department without such systems is considered to be ill managed and not under control. If teams or individuals do not comply with the system in performing their tasks they are considered to be in error. Management therefore has the task of creating the system, then of ensuring that everyone works in compliance with it. If the system is not ideal it will lead either to suboptimal work or to people working outside the system, for example by bypassing a procedure in order to make a task easier. The instinctive reaction to problems is to develop procedures to prevent their recurrence. Once procedures exist, adherence to them becomes a matter of discipline, and problems that remain can be blamed on them. It becomes difficult to improve methods, since improvement might involve noncompliance with the procedures. Nevertheless, management systems are subject to change, which varies from wholesale reorganization to minor alterations of procedures. Maintaining the system becomes a major occupation of management at all levels. And yet, if the principles are sound, we should observe uniformity and stability of such systems.

THE OPTIMUM ORGANIZATION FOR ENGINEERING

We have considered the main forms of structural organization for engineering development. In particular, the fact that the three forms, functional, project and matrix are all applied, with variations and with occasional reorganizations, shows that there is no unanimity over which form is best, and in fact managers and teachers argue the pros and cons.

Functional organization looks tidy and appeals to engineers' senses of rationality and order, as well as to their inherently scientific education. Managers of functions like it because it gives them authority and status and control over the people in their departments, and they feel that they are the best qualified to impose the standards appropriate to their functional specializations.

Project based organizations are preferred by many project managers, who would prefer to have control over all the people contributing to their project so that they can integrate the work and develop their teams. They recognize that this might be at the expense of maintaining functional standards, and that there might be problems of balancing the contributions of different specializations during the project phases and between projects. Nevertheless, particularly as a result of the teaching of Drucker, Deming, and other modern writers, project based organizations are increasingly common, particularly in newer industries or in those such as automotive that face severe competition.

Matrix management, in its various forms, is an attempt to compromise between the two extremes of functional and project management. It is therefore quite commonly applied, since it appeases both the managers of functions and of projects.

Drucker emphasized that the form of organization must be based on the nature of the business, and that no "standard models" should be used. The optimum organization for a particular business should be determined by analyzing the nature and purpose of the business, including the types of decisions that must be taken at various levels and times, and the types of relationships that must be maintained between parts of the business. He stated that integration along product lines is the more productive and effective approach, and that functional organization entails severe disadvantages in terms of generating teamwork and motivation. However, project based organization can seldom be applied totally, so that some compromise is nearly always necessary.

The optimum organization must also, of course, be based upon the nature of the people who will work within it and who will make it work. The organization will achieve nothing by itself: it is only a drawing. As explained in Chapter 3 the people who comprise the real system are not components amenable to simple rational analysis. The optimum organization must respect the facts that people have varying skills and motivations, which can best be directed towards the objectives of the business when people are organized and managed as teams whose objectives are aligned with those of the business.

To summarize: the fundamental principle of organization is that the objectives of the business, of the groups within the business, and of the teams and individuals, must all be aligned to the maximum extent practicable. Any deviation from this principle must be carefully considered and justified.

The objective of an engineering business is to stay in business. If the business is, say, the provision of test instruments, then the organization must reflect that objective. As far as practicable, as a first priority, all of the people in the business must know that their individual objectives, and those of the teams they work in, is the provision of test instruments. The objective of the business is not the design of new circuits or the buying of power supply modules: these are activities, not objectives. Therefore it is essential that such a business is organized on a project basis, with as many workers as practicable being identified with the development, manufacture, sale, and support of test instruments. Such a business will almost certainly be in competition with others, which will be working to provide better instruments sooner and cheaper. The market will be taken by the businesses that are the most responsive to customer needs for performance, price and support. Project teams are inherently much better able to respond to such challenges than are functional organizations. All of the essential criteria for individual and team performance, in innovation and in productivity, dictate that well managed project teams will outperform any other type of organization when faced with competitive and developmental challenges, as is usually the case in engineering. Therefore, in this type of business there can be no compromise with the principle of project organization, in which integrated teams are created, trained, encouraged and led, and which can operate flexibly and with the maximum of internal cooperation.

If the business develops and markets a range of instruments of similar technology, and, say, they all use similar power supplies, it might make sense to consider using a common module. Similar considerations might apply to the design of display panels or data acquisition electronics or to software development. Functionalism creeps in, and if not checked can gradually take over. The task of higher management is to ensure that the project teams retain the freedom they need, whilst being fully aware of the total effects of their decisions, in terms of cost saving to the business and of customer needs for the new product. In other words, the teams should not be tightly constrained by management, but should be required to consider the opportunities and risks and to determine their own constraints. They must be made to take responsibility for all the effects of engineering decisions on their project: in other words, for its success or failure as a contribution to the business.

Some constraints must be set by higher management. Some of these will be dictated by the market or by other external requirements, for example electronic systems must comply with standards for safety and electromagnetic compatibility. In cases such as these, higher management must ensure that individual team managers are aware and comply. Other constraints might be set internally. For example, company strategy might dictate the use of certain components or manufacturing lines. Projects must be aware of these constraints also, but should be entitled and encouraged to query them. When such constraints have adverse effects in relation to other features, higher management must make sure that the team understands and accepts the need for the constraint, or they must review the need for the constraint.

A major part of such an engineering business is often the production activity. Depending upon the type of product and the range, the number to be made and other factors, it might not be cost effective to have separate production teams and facilities for each instrument. Likewise, facilities such as environmental test and activities such as marketing and support are usually made responsible for supporting all projects within the business. The functional form of such activities is, however, consistent with the objective of the business: providing instruments. In these activities individual effort and teamwork are not directed at solving the problems of particular new products, but at manufacturing or supporting what has been developed, and there is therefore no clash of priorities between projects or with the objectives of the business.

The previous example of a company involved in the design, manufacture and support of a product range, whilst very common, is of course not the only type of engineering business. At one end of the spectrum is the small, single product business, in which the whole business is a project team, and no other form of organization is relevant. At the other extreme are businesses that provide only specialist functions, such as electronics manufacture, environmental test or fleet maintenance. Such businesses cannot be organized on project lines, since the whole objective is functional, not product related. However, managers in businesses such as these should set up project teams whenever opportunities exist, for example if new methods are being studied. The functional organization of businesses such as these must not be allowed to stifle the innate power of project teams to question the status quo and to apply inductive and deductive thinking to generate improvements.

Where does matrix management fit into this analysis? The answer is that it does not fit at all. Matrix type organizations are never optimal. They compromise the essential requirements of alignment with business objectives and with the psychology of people at work. They impose divided loyalties, between project and functional responsibilities. Does the engineer try to please his boss the project manager, or his functional boss? How is his performance appraised, and how are priorities between projects balanced? Matrix management confuses all of these essential aspects of managing projects and people in an attempt to make the management process easier and tidier. This attempt is misguided: the uncertainty and change inherent in most engineering work dictate that the organization must not be rigidly structured to ease managers' tasks, but must be kept flexible and, as far as practicable, project based, maintaining the principles of alignment with business objectives and human factors.

PROJECT TEAMS

Project teams must include all of the people who will influence the project's success. Since success depends upon performance, appearance, timing, marketing, cost and support, all of the people who can influence these must be involved at the times when they can be effective in optimizing these factors, separately and in combination This means that production and support engineers must be part of the project team when early decisions are being made on design features that can affect these downstream activities and their costs, and these engineers must have as much influence as those concerned with performance or any other features.

The composition of the team will change as the project proceeds through the phases of initial study, to detail design, development testing, manufacture and support. These phases often overlap, particularly when there is continual evolutionary development. In the early stages the project team will include mainly system designers and others who will influence major tradeoffs and factors that must be settled early, such as market sector to be addressed, capital equipment requirements and make/buy decisions. Higher management should be closely involved, but in a supporting role, avoiding the temptation to become involved in details that do not affect strategy. As the project proceeds, more people will be needed on the team to deal with the downstream activities, and some of the specialists involved earlier, for example styling engineers, might move to other work. However, the underlying principles of successful team performance are that the team must retain its identity and its leadership, and that as far as practicable engineers should remain on the team as long as their work influences that of others. The latter principle means that, for example, a circuit designer should be involved in the development testing of his design, in its production planning and early production, and in dealing with support problems. He will have been helped, of course, by having development, production, and support engineers on the project team when the design was being created.

It might appear that such an approach would be very expensive in specialist people, particularly if several projects are being developed and supported simultaneously. It might also appear to be difficult to control the various specialist contributions during the different project phases. Of course these difficulties are often quoted as justifying functional or matrix management. However, it is both feasible and practicable to overcome these problems. The engineers involved must not be constrained by narrow specializations when they can contribute over wider areas. Therefore there need not be separate engineers dealing with design, test, production planning, etc. but one or a few who stay with the design through these phases. There is no better way of making a designer aware of test and manufacturing problems than by making him personally responsible for them. There are not many engineers who are so specialized that they are not competent to do so, nor should there be. In this way the number of engineers on the project team can be greatly reduced and effectiveness greatly increased, since there is less compartmentalization and fewer communication channels. The problem of planning and scheduling the various contributions is simultaneously minimized.

The project manager must be given both responsibility for the project and authority to take decisions that affect it. This authority must include that of selecting his team and deciding what work they will do and when. Higher management or other functions should not determine these aspects, though of course they should advise on selection when appropriate, for example whether the people needed are already available within the organization or must be found elsewhere.

This approach to project team formation and management has recently been called "simultaneous" or "concurrent" engineering, and it has been presented as a radical new approach. In fact it is entirely predicated by Drucker's teaching, and of course it is supported by several recent stories of successful application. Nevertheless, the attitudes and traditions of management often lead to such successful teams being viewed with jealousy and suspicion, and they not infrequently become absorbed back into the larger organization or otherwise dispersed at the end of the project. Higher management has to impose the project team approach as the only way, and must foster and develop the concept by selection of team leaders, by giving them the necessary responsibility and authority, and by training.

INTERFACING PROJECT AND FUNCTIONAL MANAGEMENT

Since most engineering businesses need to operate with both project based and functional management, it is important that effective communication exists between projects and functions, and that the functions support both the overall business and projects. Leaving aside non-engineering functions such as accounting, the engineering support functions (typically environmental test, manufacturing, quality assurance and product support) must contribute to the projects, whilst maintaining the capability to support them by providing skills and resources. There are two main ways by which this can be achieved.

First, all functions must contribute directly to project development by being involved in early concept formulation, design and test. Of course the ideal way to achieve this is to assign engineers from the functions to the project teams, for as long as they can contribute, subject to their acceptance by the project manager. While on the team they are managed completely by the project manager, who may use their skills regardless of functional responsibilities. For example, if a test engineer can improve a design he should be enabled to do so, in cooperation with the other designers. In some cases a functional specialist might support more than one project team, but as far as practicable the principle should be applied. In this way the functions become very closely associated with the projects, and share the project teams' motivation and enthusiasm. The functional managers must encourage this, for example by encouraging the engineers to be located with the project teams and by allowing the project managers to take full responsibility for the work of all of the engineers working on projects.

Secondly, engineers should be given experience in both functional and project roles as part of their development. The amount of time in these roles should depend upon factors such as experience and particular skills, as well as the immediate requirements of the work. It should be routine for all new starters to spend time in several functions and on at least one project before settling into longer assignments. At the other extreme some engineers might have specific specialist skills that must be applied as high priority. Managers must balance these conflicting requirements, and must ensure that the long-term benefit of the organization and of the individuals receives adequate attention. Engineers can become bored and their skills can atrophy if they are kept working in narrow specializations for long periods. Also, it is important to ensure that as the needs for particular skills change with developments in technology and in markets, the engineers in the organization can respond. Of course continuation training in the appropriate skills is a necessary part of this wide development.

The managers of specialist functions naturally want to expand their responsibilities and status. They also naturally and laudably usually consider that they can make a better contribution to projects if they control the specialist contributions, for example electronics design or reliability testing. However, whilst this might be the case, and the idea seems plausible, rational and "scientific", they are not in positions where they can manage the integration of tasks and the team approach, and these aspects are much more important than the simple sum of individual specialist contributions. Therefore higher management must restrain the tendency of functional managers to build empires by undertaking project tasks.

BUSINESS PROCESSES

The term "business process" has come into vogue to include all of the various activities that are entailed in running a business. Apart from the engineering activities that we have discussed, they include processes such as ordering, supply management, invoicing, accounts, marketing, personnel, training, etc. They will all be described by appropriate written procedures. Obviously, it is important that the separate processes are efficient and that they complement one another to maximise overall business efficiency. For example, the warranty management system should use the same data and definitions on parts and failures as the design and manufacturing systems, and it should also interface easily with the supply chain management system. Therefore efforts should be made to streamline and harmonise all processes, especially where they interact. This can be achieved by *ad hoc* teams made up of people from the different functions concerned. Company suggestion schemes can also provide valuable inputs, as can management audits of how the processes are described in procedures and how they are implemented by staff.

Nowadays most business processes are controlled by the use of appropriate software. There is an increasing trend for integration of process management software within total business systems, and this imposes harmonisation and efficiency. We will look at this aspect in more detail below.

Business Process re-Engineering

"Business process re-engineering" (BPR) was an approach to optimising business processes that became fashionable in the 1990's. The approach was pioneered by the authors James Champy and Michael Hammer in their book *Re-engineering the Corporation* (Harper-Collins 1993). As presented, and as subsequently practised by the multitude of consultants who became "experts" within months of the book's publication, BPR is "the fundamental rethinking and radical development of business processes to achieve dramatic improvements in critical, contemporary measures of performance". The authors claimed that the concept represented a "conceptually new business model", and that they were "reversing the industrial revolution". Furthermore, the alternative to the application of BPR was for "corporate America to close its doors and go out of business".

These are strong claims indeed. Corporate America, as well as businesses around the world, embraced the BPR revolution with fervour. There was an enormous growth, starting from zero, in the expertise offered by the management consulting industry since the publication of the book, and the fact that over two million copies were bought attests further to the fact that the ideas took a firm hold on management thinking. Nevertheless, in the period since its inception, surveys and articles have tried to explain why BPR efforts often failed, and competing experts disagreed on the principles. What is then wrong with a concept that generated such widespread acceptance and application, and how does it relate to the management principles represented by Taylor and Drucker? What can we learn from the phenomenon?

BPR, despite its title, was not concerned principally with engineering. The process reviews were to be conducted by full-time teams (headed, according to the authors, by "Czars"), who used brainstorming methods to invent radically new methods to replace previous processes. The new processes were then imposed by management.

Of course there are organizations that need changing, and every process can be improved by incremental change or, if the process contains serious flaws, by drastic overhaul. Therefore successes were inevitably claimed for BPR, particularly by those who sold the concept. However, a large proportion of BPR implementations failed to achieve the planned breakthroughs. This was inevitable, since most existing processes in well-managed businesses will not be improved by "radical redesign". More importantly, however, humans do not respond in ways that are "scientifically" predictable: improving the system does not necessarily improve the performance of its human components. People are likely to react in ways that seem perverse to the management professionals who expect rational and mechanistic adjustment to the new order. Commonsense observation, experience and understanding of people at work indicates that change is most effective, particularly in the long term, when it is driven from the front, when everyone is involved, and when it is progressively implemented and supported, particularly by training.

Let us compare the BPR approach with the methods that have been at the heart of real business success over the last 50 years. Both Drucker and Deming taught that improvements must be led from the front (or from the top, in more conventional business terms), and that they must be consistent with and part of a total strategy. They also taught that, as far as practicable, they should be developed by or in cooperation with the people who will be expected to work the processes. BPR, however, involved considerable use of external effort (hence the sudden appearance of many consultants). The "improvements" were not led by managers, nor did staff contribute to their generation. Furthermore, the BPR approach was "radical": it involved considerable and dramatic changes over a wide range of processes, simultaneously executed.

Engineers can be rightfully sceptical of an approach that takes the word "engineering" into the context of business processes as diverse as insurance settlements and pizza delivery. More importantly, engineers know that new designs of complex human and organizational systems, especially when created by people of little experience and without the responsibility of making them work, do not work. Engineers understand the need for complex systems to be tested before they are ready for manufacture and service. Modern business processes, as we have considered above, are complex and are the results of human thought and effort, and are therefore prone to errors and oversights. That is why good business processes are developed over time, by people who know the business, have to operate the processes, and bear responsibility for their effectiveness. Of course there is scope for the occasional inductive breakthrough idea, but the world's best companies demonstrate that steady improvement, well managed across a broad front, is the most effective way to get ahead and stay ahead. This approach has the great advantage that, being deductive, everyone can contribute. Inductive breakthroughs cannot often be made to order, even by the keenest teams of re-engineering experts.

There is a further aspect of the BPR revolution that was probably even more disturbing. Not only was it based upon "scientific" thinking, it had metaphysical, almost quasi-religious, overtones. The vaunted claims made by its proponents seemed designed to frighten rather than to inspire. Did corporate America, or individual companies, really "close their doors and go out of business" if they did not accept the party line of this cultural revolution? How did excellent companies like Hewlett-Packard, ICL and Toyota manage before BPR, and did they die if they did not embrace it? There can be little doubt that the explosive growth of BPR was generated to a large extent by the exaggerated and fervid way in which it was proselytised by its gurus and their acolytes

BPR had a controversial history. Inevitably, the potentially adverse effects described earlier were inflicted upon many companies that were seduced by the approach. Some practitioners developed the concept to take more account of the negative aspects, by emphasising the need for more involvement of the people who will be affected, or by tempering the radicalism. The authors of the idea themselves later acknowledged that they had underestimated the importance of the human aspects (and, of course, their subsequent books also made money; they had become gurus). However, in the meantime great damage was done, and many companies had to recover painfully from the effects. The fad blew itself out.

The moral of the BPR story is clear: it was a reversion to "scientific" management, and was therefore in conflict with the philosophy of the new management. It should be avoided.

TECHNOLOGY IMPACT ON ORGANIZATION AND PROCESSES

Computer aided Engineering

With the rapid advances in the technologies of design and in components, materials and processes, it can be difficult for project teams to keep up to date, particularly as they will be responding to project priorities. Also, some of the technologies, particularly computer aided engineering (CAE), involve high capital and training investments that need to be coordinated. Some of the CAE tools are also very specialized, for example electronic circuit simulation and mechanical stress analysis. All of these factors can strengthen the arguments in favour of functional organization, or at least towards larger functional teams in support of projects.

The arguments for more functional specialisation must be countered by ensuring that the engineers working on projects are trained in the use of the specialised design and other tools, and that expertise is available to support the training with further help and problem solving. Some central control is necessary to provide the service, training and other support, but this must not take over the role of product design. To a large extent coordination can be managed by user groups, properly led by an appropriate senior manager such as the head of the major project design team or the director of engineering. The CAE systems must be seen as tools to assist projects, not as systems around which empires are built. Adequate training is of paramount importance to ensure that the tools do not carry an aura of mystique, and that they are used effectively.

Modem CAE, used properly, can in fact strengthen the integration of engineering effort. Modern design systems incorporate wide ranges of analyses, involving performance, detail design, and manufacturing. For example, electronic CAE ("electronic design automation" (EDA)) systems include circuit design, timing analysis, variation analysis, electromagnetic compatibility analysis, testability analysis, layouts, and manufacturing outputs to drive automatic assembly systems. Increasingly design systems are also becoming multidisciplinary, integrating mechanical, electronic, and other aspects. To maximize the effectiveness of such design tools it is essential that they are used by the team, not just by specialist designers. The initial inputs and analyses are performed by designers, but downstream functions can use the CAE tools only if the initial work has been performed with them and if the whole team is trained in their use.

Business Systems

As mentioned above, software for managing business processes has been developed to integrate a wide variety of activities. For example, Oracle™ (Oracle Corp.), SAP™ (SAP AG) and Clarity™ (Niku Corp.) are relational database systems that automate and integrate processes such as resource planning ("manufacturing resource planning" (MRP); "enterprise resource planning" (ERP)), customer relationship management (CRM), product lifecycle management (PLM), project management, financials, personnel, quality, etc. They also integrate with CAE design systems to enable transfer of design and manufacturing data between teams, functions and suppliers. The systems are modular, and versions are tailored for specific industries, so businesses can select as appropriate.

Business system software can be very effective in controlling and improving processes and in data presentation and communication. It imposes consistent data and methods and creates seamless interfaces between activities. It can also automate many tasks, such as ordering and invoicing. However, these advantages can be attained only if the people using the system are properly trained.

The Internet and E-mail

The Internet has opened up a new world of information and communication. For example, a few clicks on the mouse enables us to explore the offerings of any of the companies mentioned in the section above. Using the Internet (or company intranets) we can obtain product data, explore research results, and transfer design or test information.

E-mail speeds communication.

So what are the problems? The main one is probably that there is just so much information available, so easily, that people can swamp themselves (so-called "information overload"). In the past people had to be selective in what they sought, for example in journals, library searches, phone calls, etc., so they received what they needed over a time in which they could absorb and use it. Now everything is available immediately. By itself this is not a bad thing, but it does mean that people must be careful to seek only what they need, and not be seduced down interesting but wasteful routes. (Once on the Internet, pornography, gambling, and other tempting sites are just clicks away. Companies must impose disciplines on Internet use and have blocking systems in place to prevent misuse).

As far as e-mail is concerned, it is just too easy and very commonplace for people to find on logging on that they have received dozens or hundreds of messages, typically ten or more times more than in the days of paper mail. Dealing with so much incoming mail has become a severe burden for most managers. A large proportion of e-mail messages are unimportant or irrelevant. Forwarding the latest joke to friends and colleagues is easy and amusing, but the cost can be high. Managers, from the top down, should impose disciplines on the use of e-mail to ensure that all messages are necessary and are sent only to people who need them.

A general problem that has been created by the Internet and e-mail is that people at work can become isolated from personal contacts, spending most of the day staring at screens. Managers should "walk and talk" to keep in touch with team members, monitor their work and help with problems. Team members should also be encouraged to move around occasionally and discuss their work with colleagues face to face. Properly planned meetings can also break the monotony and stimulate the flow of ideas and develop teamwork.

CONSULTANTS

Consultants are available to advise on any aspect of business. Large consultancy organizations, often outgrowths of accountancy firms, will provide advice on everything from staff appraisal to project planning. Software companies such as IBM, SAP and Oracle provide consulting on business systems and on wider issues. Smaller groups and individual consultants specialize in different topics, such as those mentioned as well as on engineering subjects such as bearing design and electromagnetic interference. Many consultants offer training as well as advice, and some will take over management or design tasks. With the recent growth of interest in quality, and particularly the spawning of ISO9000 (discussed in Chapter 10), there has been a great increase in the number of consultants specializing in quality management and methods.

Consultants can be useful as occasional advisers to managers in areas where they know that they and their staff lack necessary knowledge. However, organizations should possess the basic knowledge necessary for their progress and survival. Managers should ensure that they and their people possess the knowledge needed, and only exceptionally should external help be sought. Many managers, particularly at high levels, instinctively react to problems by calling in consultants, whose response is often to tap the knowledge of people in the organization and then to present it to the managers as if it were the consultants' own work. Consultants often sell particular ideas, for example methods for personnel appraisal and selection, and project management methods that might not be ideal for the client. They usually sell ideas that seem highly rational and plausible, which of course is the claim of scientific management.

Consultants are expensive. However, there are other costs that are not usually identified, particularly the time spent by staff in providing information to the consultants and in discussions on the recommendations. When consultants are involved without the goodwill of the people whose work they will influence there are further large unquantifiable costs of lowered morale and suspicion. A good test of whether consultants should be called in is to determine whether the people who will be affected believe that they and their managers have, or should have, the knowledge being sought from the consultants.

For example, if an engineering company calls in consultants to advise on how to control project development, it is justifiable to question the competence of the managers who have been appointed, since managing projects should be their forte. If an electronic designer in the same company was found to be incompetent at circuit design he would be moved or removed, and the person who appointed him would have to take care not to repeat the error. The same principle must be applied to managers. An organization might be able to carry a few people at lower levels who are less than proficient, and can work to improve them. However, managers must be expected to be competent at managing their functions and teams, with the support of the people they appoint and train. Reaching for help from consultants in the primary tasks he is supposed to perform is an indication that a manager is not up to the job and has not absorbed the necessary knowledge by self-study and experience. Worse, it can indicate that he is ignorant of the knowledge possessed by his team, or even distrusts them.

If an organization is embarking on a new direction, for which past experience has not equipped it with the necessary knowledge, then it can be appropriate to use consultants, particularly in specialist areas. For example, a company might be planning to switch from small batch production to large-scale production to exploit the success of a new product and may want to use the latest methods for production quality control and just-in-time (JIT) production. If the production team lacks this experience, they should together agree to seek suitable experienced consultants.

Obviously there are shades of grey in deciding whether to use consultants. One absolute criterion, however, must be that the consultants are competent. There are many "panacea" consultants who will take on any work over the range in which they claim to be competent. They then appoint staff to the project who might lack the depth of experience necessary for the task. Consultants might also be tempted to make jobs seem larger than necessary: they are not as motivated as are those within the organization to minimise the time and effort spent on a problem.

Excellent and often indispensable help can be obtained from consultants. However, it is always preferable for the skills and knowledge relevant to problems faced to be available within the organization. When a new problem arises or can be foreseen, it should usually be possible to assign it to a manager or engineer with the brief to learn what is necessary. Information is available from books, journals, professional institutions, trade associations and the experience of others, particularly suppliers and customers. A person or team given such a task will quickly obtain the information necessary and will be enthusiastic to succeed. They will be available to continue with the project and to help on future similar problems. The knowledge will be made available within the organization, at less cost and probably more effectively than by using consultants. The person assigned to the problem should, of course, be permitted to recommend or engage consultants if he considers it necessary, but they would then report to him, not to his superiors.

RESEARCH

It has been said that the science of today is the engineering of tomorrow. There are many striking examples of the truth of this. James Watt's observations of the power of steam lifting the kettle lid led to the steam engine. Maxwell's equations linking electrical and magnetic phenomena led to radio and radar. Materials science led to the transistor, the microprocessor and all of the microelectronics components that make possible computers, modern communications, and a host of other products and systems. Nuclear science led to the development of weapons, power stations, magnetic resonance scanners and radioisotopes for materials testing. Studies of light coherence led to the laser, now used in data recording and transmission, metal cutting and precise distance measurement. Scientific research continues to present new engineering possibilities, and many companies and other organizations, particularly universities, are pursuing such "directed" research. In this section we consider how engineering-related research should be managed.

A broad spectrum exists between research that is entirely theoretical and scientific, such as particle physics theory and experimentation, and research aimed at particular real applications, such as fracture behaviour of composite materials or developing new types of electronic component packaging. At the theoretical, "blue skies" end, the future outcomes are uncertain, in terms of possibilities, benefits, time-scales and costs. At the other, "green pastures", end, the outcomes are usually much clearer, and the work can be managed to a large extent as engineering projects, with budgets, time-scales and objectives. Few companies engage in research at the "blue skies" end of the spectrum, because of the uncertainty involved. As research approaches the "green pastures" end, companies must take much more notice, and must consider at what point and to what extent they should conduct the work or participate with others already involved*. Companies should always be aware of research trends and developments that might affect their businesses. They should also be aware of their competitors' interest in such research. Research can provide the strategic impulse towards new business, or for survival. However, research is expensive and the payoffs can seldom be predicted with confidence. Therefore managing research and transferring it into methods and products is a particularly challenging aspect of engineering management.

Research is, of course, highly dependent on scientific, inductive thinking, as discussed in Chapter 2. Some aspects require teamwork, but most inductive work is individual. Therefore research management must foster both aspects, as described in Chapter 4. The balance must relate to the nature of the problem. An individual might be left alone to develop an idea, or a team might be formed to test it. Both individuals and teams must be encouraged and supported, but not forced or constrained more than absolutely necessary. Deadlines must not be set, but objectives should be agreed when appropriate. Failure must never be cause for blame, though there should then be a frank review to identify the reasons and to point the direction for future work.

Research should always be directed towards the core technologies that have been identified. A company's core technologies are those that it has determined are crucial to its future, and which distinguish it from its competitors and suppliers. How core technologies should be determined will be discussed in Chapter 6. Once this crucial determination has been performed, the strategy for research becomes

---

* An excellent example is the Dyson vortex vacuum cleaner. No new science was involved, but considerable directed research was necessary to turn the idea into a winning product. The story is told in James Dyson's book *Against the Odds*. It contains many examples of applications of the principles and methods described herein, but one point with which I disagree: he somewhat disparaged the importance of obtaining scientific understanding of the process involved. His company has moved on, and now research is given the priority it deserves.

clear: the company must maintain its strengths in the technologies identified, and its research effort must then be directed at ensuring this.

Research can be conducted internally or externally, or in combination. Internal research provides greater assurance of secrecy and allows total control of the work and of the people engaged on it. On the other hand it is expensive. Research can also be conducted externally, at universities or in research associations. External research is often shared with other companies and organizations, including sometimes government sources that provide funding and which share in the results. The results of shared research are usually published, so secrecy is then not possible.

It can be more difficult to control the direction of shared research, particularly when performed by universities. To a large extent university research work must fit in with the constraints of academia. On the other hand, careful selection of universities and researchers can provide rich rewards in terms of knowledge and facilities, and at much lower cost than internally conducted research. Further potential benefits can be derived by research collaboration with universities, particularly in relation to training of company people and recruitment.

There is a variety of ways in which universities can be employed on research, ranging from ad hoc tasks taken on by teaching staff to fully sponsored research at doctoral level. Tasks can be individually planned, or a longer-term and wider programme of work can be set up.

Research associations exist in several industrial fields, particularly in Europe. (They are less common in the United States since research shared between companies is held to be uncompetitive. However, the costs of modem research, and overseas competition, is forcing some parts of U.S. industry and government to reconsider this attitude, and the microelectronics industry has set up a joint research facility). Such associations act as research clubs, conducting both sponsored work and work of their own choosing. The results are made available to the individual or sharing sponsors or to all members as appropriate. Such associations enable smaller companies to enjoy the benefits of research in their core technologies without having to bear the full costs.

Funding for engineering research is often available from government agencies. Defence departments have been the largest subscribers to research, but this is mainly related to specific projects. Other government departments such as for industry, health, the environment and transport support research projects in their areas of interest. Within the European Union multinational schemes also exist.

Sometimes special research initiatives are set up covering particular topics, such as the British Alvey Project for information technology, the Japanese Fifth Generation Computing Project, and the European Eureka programme.[*]

The rules of the game of government-funded research are complex and variable. They nearly always require collaboration between companies and with universities. The government agencies normally provide part of the funding, typically 50%. There are usually time limits, due to constraints on government budget authorizations. Applications must be submitted explaining the work to be undertaken and the benefits, costs, organization for collaboration, and sources of additional funding. Not all applications are accepted and there can be long delays.

When government funding support for research within a company's core technology is available, it obviously makes sense to exploit it to the maximum practicable extent. Therefore the research manager needs to know what support is available and how to obtain it. Since the procedures are often fairly complex, and since there are often time limits for applications, he must be capable of putting together winning applications quickly. In every case he must have a fallback position in case the application is not supported. Generally, applications for support should not be made if the work will not proceed without it, so the application should not delay the start. Any external funding can then be considered as a bonus, which can help to speed the work or extend it.

Since research is crucial to future success, and since it is usually expensive and difficult to forecast the future benefits, it must be managed at the highest level. The chief executive must determine the strategy and ensure effective management of the programme. Depending on the amount of research undertaken, he should appoint a research committee, consisting of the managers of the functions and products most concerned, and possibly including external advisers from supporting institutions. He should also consider appointing a manager to run the research programme. The research manager should manage the people engaged on research and the progress and funding of the work His task should be to maximize the value of the research output in relation to the resources expended.

The research manager must balance the requirements for timeliness, confidentiality and research priorities in determining how work is to

---

[*] Initiatives like these have not been very successful, and, with the exception of the European Union research programmes, they seem to have lost favour.

be funded and conducted. He should maintain a network of contacts in relevant universities, research institutes, government departments and other companies, so that he has strategic and tactical information on what is going on and what is available. Such knowledge can be extremely valuable in helping to determine the research strategy.

Knowledge of patents taken out in the core technology areas, and in any other areas relevant to the business, is also important to the research programme. Inventions that result from the programme must also be patented when they provide competitive advantage or can be the source of royalty income. Therefore the research manager is an appropriate person to manage the patents search and registration function. All engineers involved in research, design and development should be kept informed of relevant new patents, and must be given instructions and help to ensure that patentable ideas are registered.

The research manager should also ensure that suitable library facilities, for books, journals and data search (including access via the Internet), are available. He should be aware of research trends with possible application to the business as well as of wider interest. Research topics that are still in the "blue skies" category might generate future strategic opportunities, and others might be close to exploitation. Research fields that hold this type of promise include superconductivity, nanotechnology, biological electronics and fuzzy logic applications.*

Organizations that exist to perform engineering research as their prime service, such as research companies, associations and the research departments of universities, must be managed according to the same principles that apply to engineering development work in terms of individual and team motivation, leadership and organization. Universities must integrate the teaching and research functions. However, a manager should be appointed to undertake the tasks described above and to maintain business contacts with industrial and government clients and funding agencies. Combining these roles with teaching usually means that both teaching and research work suffer, especially as the work involved in obtaining government funding for research can be tedious and time-consuming. A person who is free of teaching commitments, and reporting to the head of the department, can be effective in generating industrially sponsored work and research funding.

Research presents particular management challenges. Because of

---

* Written in 1993. The list still seems correct, apart from fuzzy logic. Now we could add , maybe, quantum electronics, others?

the relative uncertainty, much research has to be conducted without clear ideas of the resources needed or the results expected. All research carries more uncertainty than product or process design and development based on what is known. Therefore research and researchers must be managed with flexibility, with the minimum of constraints, but with discipline. Research depends to a large extent on inductive thought, so the people involved must not be distracted by arbitrary requirements for formal reporting or by many of the routine administrative tasks necessary for the efficient running of the organization, such as preparing budgets and liaison with funding agencies. The research manager should ensure that these tasks are performed with the minimum of interference with the time of the people conducting the research work. They should be provided with the facilities they need, including accommodation, information support and equipment without having to become unduly involved in the administrative processes.

Despite this care and attention, researchers must not be allowed to become *prima donnas* of the organization. Their work has objectives, even if these are often unclear. Engineering research is not pure research, seeking knowledge for its own sake. Decisions have to be made based on progress and on external influences, such as a change in strategy brought about by a breakthrough or by a change in market conditions. Research engineers must be subject to the disciplines of modern business. There is obviously a difficult and inherently unstable balance to be struck between freedom to perform unfettered by any constraints of time, resources, or expected results, and the sort of commercial disciplines that accountants appreciate. Few organizations manage to find this balance, or to maintain it for long. Research organizations are not immune to the growth of non-productive, peripheral functions that begin to dilute the research effort and add to costs. On the other hand, short-term accounting often results in large-scale reductions in research effort.[**]

Maintaining continuity and morale concurrently with achieving effective results and meeting long-term business objectives requires consistent vision and a high level of leadership. Research must be treated as a strategic activity, and short-term tactical matters such as one year's profit or a change in market share must not be allowed to alter the strategy.

[**] Since writing this I have seen two companies, both strongly research-dependent, eliminate their library facilities in order to cut costs.

THE ORGANIZATION OF HIGHER MANAGEMENT

The principles that must be applied to engineers in general should also be applied to higher management of engineering functions and projects. Each project must have an engineering manager. It may be appropriate to appoint a higher manager to oversee a group of projects if they are similar, or the individual project managers may report directly to an engineering director or projects director.

Engineering projects must be managed by engineers. Since the most difficult decisions and the greatest risks nearly always relate to engineering aspects, only a person who can understand these is likely to be competent to run the project. Also, it is essential for their motivation and direction that the engineers working on the project are led by a manager who can understand the nature of the problems they are working to solve.

Functional engineering departments must also, of course, be headed by people who are competent in the functions and who are respected as leaders within the specializations. They must be required to lead the whole organization in the best use of the functions they manage, so their role must extend to training and internal consultancy. They must maintain knowledge of their specializations as they are applied elsewhere, and of new developments.

The relationships between the project and functional managers is of crucial importance to the organization. Ideally most managers would have worked in both areas prior to their present positions, so they should appreciate one another's problems and priorities. The functional managers must be made aware of project priorities and must be involved in project planning so that they can plan the resources necessary to support them. Most essential, however, is the goodwill and team attitude that exists between them, and it must be a prime task of the chief executive to ensure that such cooperative relationships exist.

THE CHIEF EXECUTIVE

It is a myth of modern management that engineers are not usually best suited to the highest management position in an engineering enterprise. The justifications made for this belief are that the objective of the enterprise is considered to be primarily a business one, and that engineers are not trained and experienced in the skills of business. The business objectives and the appropriate skills are seen to be primarily financial, regardless of the product or service being offered.

Whilst it is obviously true that the business exists to make a profit and to survive as a financial entity, financing and the related functions of accountancy, contracts, marketing, deals, acquisitions and disposals are not the prime objectives. The real objective of an engineering business is to provide products or services that customers will pay for. This view of the prime objective immediately forces management to consider the long term, in terms of research, product development and staff development, which must not be sacrificed for short-term expedient. Financial management, particularly in much of the Western business culture, is overly concerned with short term figures as reflected in balance sheets and share prices.

Any engineer who rises through competence and success to a high management role can, and of course must, understand business finance and law. However, an accountant or lawyer cannot be trained to understand engineering. This is not a criticism of the intelligence or capability of accountants and lawyers, but a reflection of the reality that engineering is a science-based subject, which cannot be learned as a part-time activity. It is not necessary for the chief executive to be able to operate the bought-ledger system or to keep track of commodity prices, any more than he needs to be able to design a microwave antenna. However, he must be able to understand the processes to an extent sufficient to judge advice and to make decisions. An engineer can understand the bought-ledger system, but an accountant is unlikely to understand antenna physics. Since the risks and the major decisions, particularly the strategic ones, are primarily engineering, they must be assessed and decided by engineers.

There are well-known cases of engineers leading their companies into trouble, and they have been blamed for putting engineering before finance or for having insufficient financial acumen. Leaving aside the fact that engineers are not the only managers who have failed, the most notably successful engineering enterprises have usually been run by engineers. This is certainly true of companies that have grown rapidly as a result of exploiting new technology, and it is also true of long-established companies that retain innovative flair. All are subject to market conditions such as world politics and economics and to competition. Whilst financial experts can try to forecast the former (though for good reasons such forecasts are always highly uncertain), only engineers with experience of the product can forecast the technological opportunities and risks in relation to what the company and the competition might achieve.

MAKING THE CHANGES

Many engineering businesses today are organized on functional or matrix lines, or they create project teams that do not have sufficient authority in relation to the functions that support them. These businesses must make the changes necessary to align the organization and the individual and team objectives with those of the business. Such changes can be difficult, as indeed is any reorganization. However, changes toward project-based organization will always generate positive responses if properly planned and implemented, since they show that management thinking is directed to the success of projects rather than to the generation of internal empires, and the people involved are better motivated if they are made members of teams with clear and challenging objectives. There are likely to be some people who, for various reasons, will not see things this way, particularly those whose responsibilities or status will be diminished, and some of these might be senior and influential. However, higher management must demonstrate the courage to align the resources of the business with its objectives. Personal considerations must never be allowed to jeopardise this principle. Such considerations are temporary, whereas the basic principle concerns the whole future of the business. Also, it is essential to demonstrate that management's loyalty is to all of the people in the business, not just to a few at the top.

The people in the business are its most precious resource, so whilst jobs must not be created for personal reasons, the existence of special talents can present opportunities for the business, and jobs can be created to exploit particular talents so long as they are compatible with the overall strategy.

Having decided on what is right for the business, as a result of analyzing the objectives and strategy, the optimum organization must be decided upon. Only then should people be identified for the jobs. If the right person is not available temporary arrangements could be made, but it must be made clear that these are temporary. If people are displaced or diminished, they must be treated generously: since in most cases they will not have been deficient in their previous roles, care must be taken to explain the reasons for the reorganization and why their jobs must change or disappear, and they should be asked to suggest ideas for their future roles. They should be offered training if necessary, and any other inducements to retain their loyalty and motivation. In the last resort they must be invited to retire, on terms that recognize their past contributions. If these rules are followed the changes will be accepted, and the potential for bitterness and rancour will be minimized.

Organizational change must not be seen as a revolution that creates a new order, resistant to further change. Rather, further change must be encouraged and expected, since improvement in performance must be continuous. Also, the organization must be able to respond to variations in the business and its environment and to the opportunities generated by the people within it. As products and methods change, the effects on the optimum organization must be assessed and the necessary changes made. These changes will be beneficial to the performance of the business and to the attitudes of the people in it if they are always consistent with the basic principles set out in this chapter.

Changes must not be made without careful thinking and justification. There has been a tendency for change to be seen as a continuous necessity. The BPR craze was an extreme example, but we see job advertisements for "change managers", as though change were an end in itself, or always beneficial. We must not change what works well, unless it is clear that improvements will result. Change is always potentially unsettling and demoralizing, but if the people involved appreciate the need, the changes are carefully implemented and are seen to bring benefits, any adverse effects can be minimized.

CONCLUSIONS

The Organization as a System

Systems comprised of people at work (organizations, individuals, teams, procedures, communications, locations) have similar attributes to machine systems. For example, an engineering design system will include engineers who create designs, their managers who control them, the tools (hardware, software) that they use, design review and documentation procedures, etc. It is possible to map the overall process and to seek to optimise it in terms of economy or efficiency. If the performance of the people in the system is entirely predictable (design creation, design correctness and excellence, etc.), the system would be as predictable as a machine system. Predictability is obviously enhanced if the tasks are simple, but of course there are not many simple tasks in engineering design and development. Also, the abilities of the people involved are variable, increasingly so as the difficulties of the tasks increase. Therefore it is inevitably difficult to predict confidently how the design "system" will perform.

However, it is necessary to create and develop human systems that are efficient and predictable, even though the people involved have such variable performance.

A machine system, such as a hydraulic controller or electronic amplifier, can be robust or sensitive. A robust system is not much affected by disturbances or variation, but a sensitive system is. A control system with a high damping coefficient will be robust, but less responsive. For example, in aircraft design there is always a trade-off between stability and manoeuvrability: for an airliner stability takes priority, but a fighter must have the maximum manoeuvrability compatible with sufficient stability to prevent uncontrollable behaviour. (In fact modern fighter aircraft are designed to be inherently aerodynamically unstable, and controllability is provided artificially by electronics and software that interpret the pilot's inputs and the aircraft's behaviour. If the electronics or software fail the aircraft immediately becomes uncontrollable, and its extremely high manoeuvrability ensures that it instantly diverges from stable behaviour, even to the extent that structural failure occurs and the pilot cannot survive the stresses). On the other hand, the aircraft may adopt a stable equilibrium condition, in which manoeuvrability is zero. A "deep stall" is such a condition: the aircraft takes up an attitude in which the controls have no effect, and the aircraft descends rapidly out of control. Gravity ensures that such situations do not last very long, but other systems, such as a welding robot, could continue indefinitely generating defective car bodies if there is no check and intervention.

Control theory describes the behaviour of machine systems and enables us to predict aspects like stability, damping and control. Human systems cannot be credibly modelled in the same way, for the reasons discussed earlier. We have no scientifically based cause-and-effect relationships and values, so we cannot derive and solve equations. However, we can identify analogous behaviours that we might be able to use to design and improve human systems.

Ideally, human systems should be responsive and controllable: not unstable nor so stable that they cannot adapt to change. In addition, and unlike most machine systems, they should be self-adapting, so that the people who are in the system can themselves generate improvements. What are the factors that influence these aspects of system performance?

For maximum "manoeuvrability", the system must impose the minimum of constraints on how people work. For example, an engineering team might be created and given a very general task, such as "design and build a product that will lead the market in our field". (Of course this is a very common engineering challenge). If the team is allowed to work without having to follow set procedures or past conventions, can decide on tasks and priorities themselves without having to seek permission or approval, and do not have to report progress against detailed objectives and schedules, they will, from an external or higher-level point of view, lack stability. A common management expression for this low-stability situation is that there is a lack of "visibility". However, the team's ability to act responsively, creatively and without organizational constraints will most likely enable them to rise to the challenge and create an excellent product, on time and budget. This has been demonstrated on several well-publicised projects, including the Land Rover Discovery 4X4 vehicle and the GM Nova car. Despite these arguments and the evidence, many companies, particularly large ones, are reluctant to allow engineering teams this amount of freedom.

Principles of Organization

Like all principles of management, the first principle of organization is simple: align the organization to the objectives of the business. A secondary but important principle is to make the best use of the talents of the people who work in it. However, like other principles of management, these are fuzzy, and their application requires experience, courage, wisdom and honesty, as well as a continuous balance of the needs of the immediate future and of the long term. This is true in any business, but the organization of engineering presents greater challenges than in most other human enterprises.

Insurance companies, retail businesses, hospitals, farms, colleges and most other types of enterprise exist in relatively stable environments compared with engineering. Their "products" are not usually subject to innovative change, internally and from competitors. They are not subject to the cycle of design, development, production and support. The people involved do not require scientific education or to work in multidisciplinary teams, and the methods and processes are not subject to rapid development. Managers in some non-engineering businesses will hotly dispute these statements: for example, the development and marketing of a new food or pharmaceutical product has much in common with engineering, and innovation and teamwork can be applied to any work. There are also some engineering businesses that place less severe demands on people and management. However, most engineering businesses require a degree of flexibility and responsiveness in their organizations that surpasses what is acceptable in most other kinds of business. This makes determination of the optimum organization more difficult: in fact there can be no single, fixed optimum, since the organization affects the people in it, and they influence the organization. There can always be good reasons for exceptions, even to the extent of accepting that, in particular situations, the best person for the position of chief executive might be other than an engineer. The only people who can make the judgments necessary are the managers of the business. Outsiders such as consultants might be able to propose improvements, but they cannot be expected to know the business and its people as well as its managers should. They cannot generate the flexibility necessary for modern business. Worst, they can take away from managers the need to think for themselves and to take responsibility for their decisions.

The best measure of the effectiveness of an organization is the results it generates, both short term and long term. An excellent organization might, however, fail if the competition is even better. Therefore excellence is not enough: there must also be continuous improvement. The only way to ensure this is to create teams in which the members are enthusiastic and happy, since these are the responses that are the most sensitive indicators of good management.

6

MANAGING ENGINEERING PROJECTS

Engineering projects always involve resources of people, knowledge, facilities, time and money. In the previous chapters we have considered the management of the human resource, and in this chapter we consider further aspects related specifically to organizing projects. The material, financial and temporal resources must also be managed effectively. In this context we will include the management of suppliers to the project, that is, organizations whose people are not under the control of the managers leading the project.

Engineering projects may be led by the customer, or the customer may play no part in the project. Examples of the first situation are a power utility placing a contract for a new generating station or a government department ordering development of a new combat aircraft. The second situation is typical for projects such as domestic equipment, vehicles, electronic components and a vast range of other products. In between is a range of products for which the customer influences design and development in limited ways. For example, a vehicle manufacturer typically will work closely with suppliers of critical subsystems such as transmission components, particularly if new techniques are being applied.

In every case the resources necessary for the project must be organized. The right people and facilities must be available when needed, the timing of activities must be planned in relation to the availability of resources and to the demands of the project and the financial provisions must be budgeted.

PEOPLE

At the outset of a project the right people must be selected to perform the necessary tasks. The first choice must be the project manager, or leader. It is essential that the leader is chosen before any other selections are made since he must be closely involved in these selections, and as far as practicable he should make them.

The leader qualified to manage the project, in three major respects. He must be competent in the main engineering disciplines that will be involved and he must be competent as a leader. He must also believe in the success of the project. From the first criterion it follows that he should be an engineer with the appropriate experience*. Once a leader who has these basic capabilities has been selected and briefed, he should be trusted by higher management to make the correct decisions regarding the other resources that will be applied to the project.

The leader will usually have been involved in the early discussions and studies that lead up to the launch of the project, or he might have been the initiator, and there are obvious advantages in allowing a person with such qualifications to take the leadership role. In other cases he might be selected after higher management has decided to embark on the project.

The leader must be briefed on the major objectives and constraints. Objectives might be based on customer requirements or specifications, market studies or exploitation of an internally generated idea or invention. Constraints will be related to the overall resources available: primarily finance and time. Other constraints might also exist, such as company policy on use of capital facilities and remuneration policy, but these should support the project, not hinder it. These requirements and constraints set the boundaries, or strategy, for the project. Within these boundaries the leader must be given wide discretion to plan and execute the tactics he determines to be appropriate.

The other people needed for the project should be selected by the project leader. Of course people might be nominated from other functions or projects, but it is important that the leader accepts them, and that they know that they have been so accepted. This point is crucial to team building and performance. The people assigned must transfer their loyalty to the project, and this will happen only if they are personally interviewed and welcomed by the leader. The leader must have the right to refuse to accept people onto the team if he considers that they are not necessary or will not contribute effectively.

---

* Exceptions can be made in particular circumstances, for example, if the project's success will be heavily dependent on non-engineering factors such as design or fashion. Examples are mobile phones and domestic equipment ("white goods"). However, he must then appoint an engineering manager, and they must work as a team. James Dyson, mentioned earlier, is a good example: he is not an engineer.

The roles and tasks assigned to members of the team must also be decided by the leader and accepted by the members. It is essential that roles and tasks are not constrained by functional terms of reference and allegiances. An engineer on the project team must be employed as the needs of the project and his competencies dictate, even if this means taking on tasks outside his functional terms of reference.

The project team must consist of the minimum number of people necessary for the tasks. The tasks will of course vary as the project proceeds, from initial studies through detailed design and development to production and support. The team composition and size will therefore have to be changed in response to the tasks. Team size obviously influences costs directly. In addition, every person added to the team increases the difficulties of communication and control and adds further interfaces. Adding more people to a project team therefore does not necessarily reduce the time taken for tasks to be completed. This means that every member must be able to contribute to the maximum extent, limited only by his capabilities, and not by arbitrary terms of reference or functional responsibilities.

To minimise the time taken to perform engineering tasks such as design, planning and conducting tests and solving problems, as well as not to generate problems in the first place, the team must consist of people with as much relevant experience as possible. Also, when experience is gained by work on the project, this must be exploited as the project proceeds. Experience includes aspects such as knowing the others in the team and the status of the project, not just purely engineering aspects. The project leader must have, and use, the flexibility to make the best possible use of the talents and experience of the team members, as this is the most effective way of keeping the team small, cohesive, and productive. It also is the best way to generate enthusiasm, teamwork and pride, as well as future project leaders.

KNOWLEDGE AND CAPABILITY: CORE TECHNOLOGIES*

The knowledge a company possesses resides in its people and their experience, as retained in brains, documents and patents (often referred to as "intellectual property" (IP)). Its capability also resides in its people as well as in its other resources, such as laboratories and

---

* For an interesting example and discussion of this topic, see R. Venkatesan, *Strategic sourcing: to make or not to make*, Harvard Business Review Nov/Dec 1992.

manufacturing facilities. These are the features that determine a company's strategic position. However, these strategic components exist relative to the competition, in constantly changing patterns. People develop and move, the knowledge base expands, patents lapse, and facilities age in relation to technology. Competitors' capabilities also change. Therefore the product strategy must be part of a wider management strategy that ensures that the advantages are maintained and enhanced.

The knowledge and capability possessed by a company can be thought of as its core technologies. These must be identified and developed. Identifying the core technologies is not, however, simply a matter of listing what the company is good at or what it has done traditionally. The core technologies are the knowledge and capabilities that exist now, and which will be made to exist in the future, that give it potential or actual advantage over its competition. For example:

- The knowledge and capability to design and develop new generations of mobile phone systems is a core technology of companies like Nokia, Motorola, Sony/Ericcson, Siemens and others. However, these companies do not design the standard electronic components like resistors and memories, electrical power supplies and cooling fans that form essential parts of the systems.
- Companies that make and sell personal computers, such as Apple, Dell, HP and Sony, concentrate on system design and build. They do not design or make microprocessors, memory chips or disc drives. The development of portable computers (laptops, notebooks) provides a good example of the benefits of concentration on core technologies. In this case the differentiating features that enable a new machine to excel are the display and the extent to which the electronic circuitry can be designed to be very compact and to consume very little power, so extending battery life. Items such as the microprocessors, memories, keyboards, batteries, and plastic housings are identical or very similar in all such products. Some of the leaders in portable computer development therefore invested heavily in the core technologies, particularly in displays. As a result, these leading companies can now provide a steady stream of new products with improved features, and at the same time can sell displays to their competitors at premium prices.

- Car manufacturers do not design and make standard items like batteries, lighting and tyres. However, most design and build their own engines and powertrains.

Core technologies can be related to design or to manufacturing capability. The product strategy must distinguish between these, again taking account of the strength relative to the competition. Companies might decide to manufacture their products because they have the capability and experience, even when it might be more economical to "outsource" the work. This is most marked at the lower technology end of the market, where price competition is most acute and when several manufacturers are able to build and sell nearly identical products.

The market for a large range of electronics-based products has, in recent years, provided a good example of this situation. Virtually all electronic assemblies are based on printed circuit boards (PCBs), on which components are placed and soldered, followed by test, and the "loaded" PCBs are then assembled into the final product. Most electronic equipment manufacturers used to make, load and test their own PCBs and higher-level assemblies. However, the cost of maintaining these capabilities at a competitive level, with the advent of multi-layer, fine line PCBs, much smaller but more complex components, and automatic component placement and test has become so high that they can be justified only by high volume throughout. The work is also very specialized, though the processes are nearly identical regardless of the actual circuit function, so a manufacturing facility can relatively easily handle a wide variety of circuit types. Furthermore, the manufacturing and test processes do not normally confer strategic product advantage, except potentially in relation to cost and quality.

Several specialist electronics manufacturing companies have been set up, as spin-offs from existing companies and *ab initio*. Their core technology is electronics manufacturing. Manufacturing capability need not necessarily be considered as a core technology of the company that designs, develops and sells the product. Consequently, several companies now provide this specialist manufacturing service, and many electronic system manufacturers now subcontract out all or much of this work.

A similar situation exists in relation to what are termed application-specific integrated circuits, or ASICs. ASICs are circuits that are designed and made for a particular application, in contrast to standard integrated circuits such as memories, microprocessors or signal processing circuits. Since integrated circuit manufacture is a highly specialized operation, requiring very high and sustained capital outlay, as well as high volume throughput in order to be competitive, very few system manufacturers make their own ASICs, even if they design them.

Generally, manufacturing capability should be considered a core technology if any of the following apply:

- The capability includes important intellectual property (techniques, designs, experience) that provide competitive advantage, particularly if it is important that these are protected.
- It is important that a strong link is maintained between design and manufacture. (This aspect is discussed in Chapter 7).
- The capability is not available elsewhere at competitive costs, and from competing providers.

The manufacture of cars, involving heavy capital outlay on robotic production facilities, dedicated to a particular model, is a classic example of a core manufacturing technology. No car company outsources manufacture.

The manufacture of mobile phones presents an interesting current example: should this capability be considered a core technology by the companies that develop and sell them? The production rates involved means that for most lines the cycle time (assembly, test, packaging) is about 30-60 seconds, well beyond the capability of most general-purpose electronic production systems. Based on the three criteria listed above, such a capability would seem to represent a critical core technology in such a highly competitive industry. Despite this, at least one mobile phone company (Sony/Ericcson) has recently outsourced handset manufacture.

Make Or Buy?

The decision whether to make or buy parts of a product is closely related to the question of core technologies. For any product that will be made up of other parts or subassemblies, decisions must be made on whether to make or buy the various items that together form the final product. These decisions must also be based on the strategic assessment of what constitutes the core technologies of the company. Any component or subassembly that can be provided by multiple potential suppliers, whose competence matches or exceeds that of the final product manufacturer (usually referred to as the "original equipment manufacturer" (OEM)), should be bought, since competition between the suppliers should ensure satisfaction in terms of innovation, price, quality, and service. Only when the item concerned confers particular competitive advantage for the product should it be considered for in-house development and manufacture.

The judgment of exactly which items generate competitive advantage must be made by considering the factors that are intended to differentiate the product in the eyes of the customer. These might be particular aspects of performance, cost, quality, time to market, production volumes, and fluctuations in demand. Note that some of these factors can militate in favour of making, and some in favour of buying. They must all be considered together and projected into the future, in relation to the product being planned and other future product strategy.

Responsiveness to production demands and fluctuation is particularly important when these aspects are not easy to forecast, or when the investment necessary for production is high. It is always easier and cheaper to increase or decrease order quantities of bought-in items than of those made in-house, so long as adequate competition exists.

We have discussed some examples of make/buy in electronics above. In mechanical engineering there is a wide range of items and technologies that companies must consider. For example, an engine manufacturer must consider items such as fuel control systems, castings, pistons, piston rings, and machining operations. A manufacturer of earthmoving equipment or agricultural machinery must consider engines, power transmission systems, hydraulic systems and components, castings and forgings, and a wide range of other detail components.

The make/buy decision can often involve existing in-house capabilities. It is common for companies that have developed and manufactured particular components for their products in the past to have to face the question of retaining or closing down the capability. The criteria listed above must be applied, but there are other possibilities that should be considered.

If an existing in-house capability ranks high in relation to potential suppliers, but is not a core technology, it could be given the opportunity to compete for the business, and possibly also for external business. In order to do this, it must be given the freedom and motivation to generate improvement and innovation in order to maintain and increase its competitive position. It must not be hindered in relation to external suppliers by arbitrary overhead cost burdens: it should be allowed to negotiate with the parent company for only that part of the corporate overhead essential to its operation. A further alternative is to allow the facility to operate outside the parent company, either as a wholly-owned subsidiary or as an independent company. An example of this was the hiving-off by General Motors of the Delphi electronics business.

For items that do confer strategic advantage but which are nevertheless bought, and where aspects of design, development, or manufacture are important in relation to differentiation of the final product, it is essential that the final product manufacturer retains the capability to specify and evaluate the item. This could include aspects such as prototype manufacture and test, and the ability to test the suppliers' products. This competence must be made to operate in close partnership with the suppliers to ensure that the items are designed and developed to provide the product differentiation needed. The engineers in the buying and supplying companies must work as a team, both sides understanding the requirements, processes, capabilities and limitations of the other. Teamwork is also needed for standard items that do not provide differentiation, though the extent of engineering involvement with the suppliers' designs and processes would be reduced.

Products whose differentiation is based on non-engineering factors, such as good marketing and distribution, or on low-technology activities such as manual assembly, can of course be competitive, and again the personal computer market is a good example of this. Several companies have grown rapidly, providing machines which are so similar to the competition that they are referred to as clones, and whose only differentiation is price. Their manufacturers merely assemble them from standard components and subassemblies, and carry out very limited design and development. In other words, they have no core engineering technologies. Such operations are, however, vulnerable to developments by companies that do strive to maintain core technologies: a breakthrough that provides new features that customers want can leave the "clone-makers" stranded.

It is essential to ensure a strong link between the make/buy strategy and the research strategy. Research must be concentrated on the core technologies. The management of research was discussed in Chapter 5.

Overseas Manufacture

It is common practice for companies to move manufacturing operations to locations where costs are lower, particularly to Far Eastern countries such as China, Taiwan and Korea.* This is often a logical move. However, there are potential pitfalls and disadvantages that should always be considered. Probably the most important, particularly if all or some of the manufacturing operations are core technologies, is the increased difficulty of ensuring that the necessary close link between design and manufacture is maintained, especially during the start-up phase. Modern communications, particularly the Internet and E-mail, now ease this problem.

TIME

The time taken from the decision to proceed on an engineering project to delivering products to the market is a very important factor affecting business success. First, time must be paid for in terms of the salaries of all of the people involved, as well as overheads and other project costs. Secondly, the earlier a new product is available for sale, in relation to competitive products, the greater is its market impact and also the potential for higher prices and profits. Thirdly, time spent on delayed projects represents lost opportunity, as the people involved could be working on further developments to widen the market or on other new projects. Therefore time is a critically important resource for project management attention.

It is not practicable to forecast with certainty how long engineering tasks such as creating a new product design will take. So much depends on the skill and motivation of the people involved and the problems they encounter. There is no such thing as good luck in engineering, just varying degrees of bad luck, so no project of other than minimal complexity and risk is without problems in development. Nevertheless, it is essential to make some forecast of the time needed, and then to manage so that it is minimized.

Since time must be minimized, and since there will always be unforeseen problems in the future, it is important that tasks are not allowed to use the time allocated when they could be completed more quickly. All tasks must be expedited so that the people concerned can proceed to other work. A closely integrated team, made up of people

---

* Far Eastern companies also move production westwards: the best known examples are the Japanese car companies' moves to the USA and UK.

who are organized and willing to undertake whatever project tasks are within their competence, makes possible this flexibility of response.

Organizing and managing time in engineering development is truly difficult. Since time translates directly into money as well as being a crucial factor in competitive marketing, it must be planned and controlled. In principle, planning and control of time can seem straightforward, and standard techniques have been developed. The best known is the Programme Evaluation and Review Technique (PERT). The method is also called Critical Path Management (CPM). All PERT requires is that the durations of all activities and their relationships to one another are defined. Then it is a relatively simple task to create, manually or by computer, a chart from which can be determined the total project time and the critical paths (series of activities which must be completed on time in order to achieve the total project time). When changes occur, typically an activity taking longer than expected, then the effect on the project timing and critical path can be determined. The PERT is initiated at the start of the project and is continuously updated.

If uncertainty exists regarding activity durations this can be built in by, for example, inserting likely minimum and maximum times or time distributions. The PERT system will then analyse the effects of these uncertainties on the overall project.

PERT is used widely in engineering development. It can be a very useful and effective method for planning and for monitoring progress and for evaluating the effects of problems or changes. It can also become a monster that inhibits good management and creates large amounts of unnecessary work. It can even seem to be both at the same time: a useful tool for senior management and a time-wasting irritant to people working on the project. Senior management observes the visibility of the project and its various activities and can use the PERT outputs as cues for action. What the outputs do not show, however, is the extent to which the PERT system itself might slow the project and add cost and confusion.

Project management systems such as PERT can become detrimental to good management when they are used to forecast and control situations with large amounts of uncertainty, or when the interfaces between activities are complex. Since engineering projects are often like this, the use of PERT must be carefully judged in order to obtain the benefits without incurring the penalties.

Planning must start with the main events that are to be achieved. For example, a time to market might be identified, or a contract might have a delivery date. This will determine other key events, such as first prototype test, start and finish environmental test and production start. In turn these events will determine lower level activities, such as team buildup, start design, select components, assemble prototypes, purchase production tooling, etc. For most engineering projects a large number of activities can be identified, and there will be many interrelationships. For these to be input to a PERT system, all of the activity times and interrelationships have to be known or assumed. The problem is, of course, that there is a level below which many of the times are difficult to forecast and many interrelationships are hard to define. These conditions apply most strongly to the highest risk aspects of the project, which require the most management attention. In such areas engineering work tends to be more iterative than sequential.

Problems arise that cannot have been forecast, and therefore action on them will not have been planned. Problems in one area can influence others: for example a material problem might lead to re-design and re-test of a machined component. These types of uncertainty and complexity are largely absent from non-engineering projects such as building a house extension or baking a cake, the kinds of example often used to illustrate the usefulness of PERT. Such examples can mislead by implying that all projects can be broken down into activities with known times and simple, sequential dependencies. As we discussed at the start of Chapter 5, it is naïve to believe that engineering work can be reduced to a simple series of linked, deterministic tasks.

A major reason for the difficulty of assigning times to engineering activities is the fact that they are performed by people whose performance, and therefore the times they take to complete activities, can be very variable. Times depend upon the difficulty of the tasks, the skill and experience of the people involved and on their motivation. As explained in Chapters 3 and 4, these factors can have significant effects on performance. In fact, applying PERT to such activities can actually degrade performance. Time estimates given will tend to be "safe", and then there is little motivation to improve on them. When activities are not achieved on time, excuses will be sought among the complex interrelationships. Teamwork breaks down and managers must resort to driving large numbers of small problems. This situation can quickly escalate to the point where the project becomes a complex series of crises, with the PERT system providing a misleading aura of control.

The rationale for PERT, or for any planning at the level of detailed tasks, is based on the "scientific" approach to management, as discussed in Chapter 3. By discouraging initiative, flair and teamwork in dealing with hour-to-hour and day-to-day problems, it can lead to demotivation and reduced effectiveness in the same way as work study methods do for production tasks. In fact, the adverse results can be more dramatic, since the uncertainties and relationships in development projects can be so much greater, and the risk much higher, than in the planning and execution of production tasks.

Project time planning and control must be based upon careful judgment of the tasks involved and their relationships, and the uncertainties and possibilities. The key activities must be identified, and planning techniques such as PERT can be useful for analysis and display. However, only key activities should be analyzed in this way. Key activities are those identified as crucial to meeting the overall time constraint, and on which other important activities depend directly and unambiguously. Next, the extent of uncertainty associated with each key activity must be identified and the action needed to monitor and minimize it must be planned. For example, a design involving techniques new to the organization should be started earlier, since more iterations will be likely than for a design based upon recent experience, even if the complexity is similar. Activity duration and uncertainty estimates must take account of the skills, experience and motivation of the actual people assigned, and they must contribute to the estimates and accept the challenges. "Synthetic" times, based on past experience but taking no account of these particular realities, can be highly misleading and should be used only for low-risk activities.

By concentrating on key activities, the problem of trying to analyze and plan for complex interrelationships and iterations can be largely avoided. These are tactical aspects, which should be managed flexibly by the teams involved. In the same way that production workers should not be told in detail how to perform tasks, engineers involved in projects should not be given detailed instructions on methods, times and sequences. They should be allowed the flexibility to determine the best approaches as the project proceeds and to operate as a team. Senior managers must avoid the temptations of planning the details in advance, and then trying to "control" the many separate activities. Senior managers need to be aware of progress and of problems. They must concentrate on strategic issues, such as the provision of capital resources or changing the times of key events, but they must allow the team to handle the tactical issues, and they must trust them to report on progress and problems.

Time Management

Much has been written and taught about the importance of how individuals should plan and manage their time. Obviously, managers must make sure that, as far as is practicable, their time and that of their staff is spent productively. However, in most engineering work there is not a clear distinction between productive and non-productive time. Rather, there is a continuum.   Time spent browsing through an engineering journal or web page might yield valuable ideas or information.   Alternatively, time spent unnecessarily in a project meeting or in performing flawed design or test work is wasteful.

Ultimately, it is the productivity of the team that matters, not just that of individuals. Managers must take all appropriate steps to ensure that time is spent as productively as possible. This is not the same as micro-managing the minutes and hours of individual workers. The effective management of time requires a careful blend of trust, discipline and understanding: in a word, leadership.

Time Recording

Time recording systems are sometimes used, which require people to record the time they spend on different activities such as work on particular projects or tasks, training, "diversion", holidays, etc. This all goes into the computer, and management is provided with "visibility" of what their people have been up to.   However, as discussed in the preceding section, this can be misleading. There are some aspects of time recording that are needed for business reasons, such as sick, holiday and training times, but the system should not be used for detailed management of working time. People working on a project or in an engineering function have jobs to do and their managers should know how they spend their time and how productive they are.   Therefore, precise recording of times spent on lower-level activities does not provide further useful information and it wastes time. Time recording systems should be kept simple and should only record data that is really necessary.

FACILITIES

The facilities required for engineering project development range from simple tools and instruments which can be obtained for use specifically on the project whenever they are needed, to expensive fixed facilities such as test chambers, the use of which must be planned and scheduled. Some facilities will be required wholly for the project, and others might be shared. The project leader must organize the provision of all necessary facilities so that they are available when needed, with maximum benefit to the project but at economic cost.

A facilities plan should be prepared, defining all the items that will be required, with the times and costs. It is obviously necessary to plan and budget for expensive or long lead time facilities at the start of project planning, since these will influence overall costs and timescale. The need for such facilities can usually be identified without much uncertainty. Suitable team members must then be made responsible for their provision, if necessary working with other functions in the organization or with external suppliers. Other facilities will be harder to forecast, and requirements might change as the project proceeds, so the plan must allow for these contingencies.

Just as it is necessary for senior management to trust and support the judgment of the project leader, he must trust and support the judgments of the engineers in the project team regarding their claims for the resources they need. Whilst it might be instinctive to assume that they will ask for more than they really need, it is usually wiser to accede to requests for high quality facilities. This demonstrates trust in the engineers' judgment and enhances their motivation and esteem. If errors of judgment are made that result in too much being paid than is really necessary, say for a test instrument, this is almost always less costly to the project than an error in the other direction, leaving aside the influence on motivation and therefore on performance of the people involved. Arguments over cost reductions often cost more in time and money than acceding to requests based on trust. If mistakes are made, lessons will be learned that can be applied in the future. It is also often the case that the value of using enhanced facilities becomes apparent only after they are put to use. For example, better equipment might be more reliable and might have features that extend their versatility and useful life.[*]

Decisions on provision of facilities must be made quickly, especially when they are needed urgently. Projects must not be constrained by arbitrary rules on capital expenditure, such as budget limits, timing and authorizations. The project leader must be authorized to obtain what he needs when he needs it, subject only to his overall constraints. Enthusiasm and motivation can be quickly dissipated if the

---

[*] This discussion gives further support to the importance of engineers managing engineering projects.

management system is not seen to be supporting the project team to achieve the objectives it has been set.

Of course there are sensible economies that can be applied to the provision of facilities. Stated needs should be questioned if they appear to be over-ambitious, but the person making the request should agree with the decision. Economies can be made by sharing facilities, such as environmental chambers and test instruments, but only if sharing does not adversely affect availability when needed. An error that is sometimes made is to provide shared facilities on the basis of "average utilization", without taking account of the delays and confusion that can result from people having to wait for their turn to use an item. A common result is that people will not use the facility at all, so the expected increase in utilization does not occur and the project suffers. In fact utilization rate of such items is an irrelevant criterion in comparison with the costs of waiting time, frustration and reduced morale.

It is often practicable to economize by ensuring that everyone on the project is aware of what facilities are available. If a project facilities plan exists the team should have access to it. Facilities that are no longer needed can be declared and held in a pool either within the project or on behalf of several projects. Such an arrangement can be very effective, but only if the people working on projects know that if they declare items to be available to the pool, their needs will be satisfied either from the pool or from outside. The response to perceived under-provision of resources is for projects to hoard their assets, even when they are no longer needed.

The question of whether to buy facilities or to hire them, or, in the case of fixed facilities whether to hire time and associated support rather than build new ones, must be based on the cost differences over the period that the facilities will be required. For example, it would normally be more cost effective to hire an expensive instrument if it were required for only a few months. Hiring has other advantages: it obviously reduces capital outlay and depreciation, as well as over-heads such as calibration, maintenance and storage. It ensures that the most up-to-date equipment is available if needed and that delays due to failure are minimized. On the other hand, engineers might be less familiar with equipment that is used only for short periods.

If several projects are being developed, it makes sense to consider the provision of facilities over the range of projects, not just for each project in isolation. Of course each project must determine its own plan, but management must ensure that economic provision is made across all projects and for future ones. Hire or buy decisions must then be taken at higher levels, but the principles of support for the needs of the projects must be maintained. It is important to guard against bureaucratic functional control, which can restrict the supply of facilities by being subjected to arbitrary centralized budgets and procedures.   In multi-project organizations a manager can be appointed to provide all facilities required by projects. He should maintain a pool of items and publicize their availability. He should provide any facilities needed by projects, either from the pool or by hire or purchase. The costs of hire or purchase should be borne by the project, unless it is decided that wider capitalization is appropriate in the interests of other, or future, projects. The facilities manager should maintain good, long-term contract relationships with selected hire companies so that the facilities that are needed can be obtained quickly and at minimum cost. There are also advantages in minimizing the number of different types of similar items and numbers of suppliers. For example, different test and measurement equipments from one supplier are likely to have similar features which will enhance their productivity, and concentrating on a few suppliers can also enable better commercial terms and support to be obtained.

Once facilities have been provided as part of the capital structure, for example computer-aided engineering (CAE) facilities or fixed test facilities, their use should not be charged to projects but should be depreciated as an overhead burden. Charging these costs directly to projects can lead to unnecessary additional purchases and other expense. Also, such charging discourages the very use of the facilities that have been provided to enhance efficiency. The facilities manager must be tasked with providing the support needed by projects, while taking care of the wider and longer-term needs of the organization. He should advise on matters such as relative costs of hire and purchase, and he should provide infrastructure such as laboratory space, documentation, calibration and support.   Engineers working on projects must prefer to deal with him rather than with external suppliers. However, he must be prepared to encourage direct dealing with recommended external suppliers when appropriate in order to minimize delays and staff-work.

Some facilities justify a standardized approach by the organization as a whole. CAE systems are in this category, since they enable the transfer of engineering information between projects and functions (particularly between design and production, and between design and suppliers), and because there is a large investment in training and experience. On the other hand, since good CAE systems can greatly influence engineering productivity, the needs of the projects must be satisfied. Such key facilities as CAE, test chambers and laboratories are strategic assets, requiring top management judgement and decisions. Their provision must include all of the support necessary, particularly training, and necessary enhancements that become available to ensure that they are used effectively by projects. The productivity gains generated by such investments can be maintained and improved only by ensuring that engineers use the facilities effectively, and are encouraged to do so by training and support. Senior engineering managers must plan for future needs by providing such strategic assets and support in advance of particular project needs, since project leaders will not always have the time or knowledge needed for such judgements.

INFORMATION

Information is an important resource in engineering development. Engineers working on projects must have easy access to relevant information on materials, components and processes, and on developments in the market sector, such as on competitive products. Information can be made available by ensuring that engineers have easy access to relevant magazines and journals as well as to books and other data such as standards and catalogues. An efficient library service is an essential support facility for any engineering development work. It should include an abstracting service, tailored to the needs of the projects supported. For large organizations this can be a central service. For small organizations, or within individual projects, a local information centre which provides relevant standards, catalogues, clippings from technical journals, etc. should be set up.

Important or useful new information should be circulated to engineers by name, and a regular circulation folder containing such information is a good way of achieving this. Engineers should be encouraged to contribute items to the information centre and the circulation folder.

This kind of information can now be accessed on the Internet, and to some extent the Internet replaces the need for some of the services described above. However, managers must ensure that their people are aware of what they need to know and do make appropriate use of it. For example, if the design must comply with a particular standard the relevant engineers must know about it. The library and the local information centre could include and disseminate information about relevant websites. Also, an information service could be provided via a dedicated library or project site.

SUPPLIERS

External suppliers to projects, whether they provide services, hardware or software, must also be organized and controlled. A typical project might require the use of external test services, design and manufacture of subsystems, both hardware and software, and provision of subassemblies and components.

Suppliers, whether major subcontractors or providers of components or simple services, are important contributors to the success or failure of the project. For example, on typical systems a large proportion of all failures in service, and therefore of warranty and other costs, are "bought" from suppliers. On large or complex systems this contribution can be as high as 80%. Therefore suppliers must be selected on the basis of their capability and their commitment. They should be briefed as carefully as internal project staff, and they should be prepared to accept responsibility for their contributions in terms of performance, quality, delivery, cost and any other factors that can influence the project. Of course more attention must be paid to a supplier providing a complex subsystem than to one providing standard components, but the principles should be the same: there are innumerable tales of project problems generated by insufficient attention to the details of applications of "simple" components*. Good suppliers are eager to support users of their products and services, and this support is often without cost. For example, most suppliers of engineering components and materials provide free applications advice, in addition to data and handbooks. Any supplier reluctant or unable to provide such support should be used only as a last resort, and his contribution should be very carefully managed. Suppliers must not be selected on the basis of price alone. The practice of selection based on lowest cost tendering is very poor engineering, and not very good short-term economics. Initial cost is only one of many factors to be considered, albeit often the easiest to quantify and with the greatest short-term effect. However, the other factors are usually much more important in relation to the success of the project over its total development, production, and use. Therefore suppliers should be selected by the project leader, not by separate tender boards or purchasing departments.

---

* One glaring example: a manufacturer of agricultural machinery used cheap, low-cost electrical connectors on a new machine that, for the first time in their experience, used electronic control systems. The connectors could not withstand the very severe environmental conditions of vibration and dust.

As far as practicable, particularly when support is required over a long period or for several projects, long-term partnerships should be developed with key suppliers. These partnerships should mirror the philosophy applied to the selection of key personnel within the project organization: trust and commitment. In addition to the obvious economies in terms of issuing invitations to tender and reviewing responses, this approach can result in better communication and much stronger commitment to the project by the chosen suppliers. Of course the selected suppliers must be subject to review, and special circumstances might dictate using alternative suppliers, so the approach must not be applied rigidly, but the chosen suppliers must know that they will be preferred so long as their performance matches the demands placed upon them.

Suppliers should not be selected on the basis of accreditation or approval by other organizations, unless this is an essential requirement for contractual, safety, or other overriding reasons. In particular, generic "systems" standards such as ISO9000, the international standard for quality systems, or related national standards, which involve assessment by third parties of organizational and procedural aspects, should not be relied upon to provide assurance of quality or performance. These standards usually impose only very basic requirements, and the third-party assessors cannot evaluate the products or services in relation to particular applications. They are discussed in more detail in Chapter 10.

PURCHASER-SUPPLIER RELATIONSHIPS

On many engineering projects close relationships exist between the purchaser and the supplier, with engineers on both sides being involved in specifications, design and development activities and support. Such projects are usually large, and often involve several layers of subcontractors and suppliers. Typical situations are power generation plant for electric utilities, chemical process plant for large chemical companies, train systems for railways, aircraft for airlines and missile systems for governments. The nature of the contracting relationship between the purchaser and the supplier can greatly influence the outcomes of such projects in terms of cost and performance.

At one extreme the supplier decides what is to be designed, developed and made, and sells this product to the purchasers. Of course the supplier will take account of purchaser requirements, particularly during the early concept formulation phase, and he will probably discuss these requirements and react to them during later development, or even after the start of production. We see examples of this approach in the way aircraft manufacturers develop airliners. There is close contact with the major airline customers whose requirements influence the designs, but the airlines do not write detailed specifications, operate development contracts or make decisions on the project. At the other extreme, a government defence procurement agency might spend considerable effort in determining detailed requirements and specifications, which are then passed to the supplier via contracts that stipulate methods and standards to be used, progress monitoring by the purchaser, payment arrangements and other aspects. Some examples will help to identify the strengths and weaknesses of different purchaser-supplier relationships.

- British European Airways, then the nationalized British airline covering the European market, specified their requirements for a new jet airliner to replace the turbo-prop aircraft in service during the 1950s. The requirement was placed with Hawker Siddeley Aviation, who questioned the size of aircraft specified. The manufacturer considered that a larger aircraft would be more appropriate for the airline, and would be easier to sell to other airlines. However, BEA, with government support, forced the development of the aircraft that became the Trident. Even before the aircraft entered scheduled service it was apparent that it was too small, and BEA were eventually forced to buy larger Boeing aircraft. In the meantime, Boeing developed the larger 727 three-jet airliner, which became the largest selling jetliner in the world, whilst the Trident was a total failure in overseas markets.

- During the 1950s the US Department of Defense worked with defence equipment companies to develop an anti-aircraft missile system. The requirements were specified in detail, and it was intended to satisfy these by using the maximum of existing, proven sub-systems such as radars, tracking units, computers, etc. The Mauler system was developed, but could not be made to achieve its performance requirements and was eventually cancelled. At about the same time, the then British Aircraft Corporation embarked on a privately funded project to develop a system to meet the same general requirements. However, BAC wrote the specification and retained complete freedom of action on design and development. The resulting system, Rapier, has been highly successful, and developed variants are still in use.

- In the 1980's the UK government initiated a project to create a standardized "military microprocessor". It had been decided that the then commercial offerings would not be good enough for military applications. The specification was written and the development contract was awarded to Ferranti Ltd. Millions of pounds were spent, but the project was rapidly overtaken by the rapid evolution of commercial products, particularly from Intel and Motorola, no useful hardware was created, and it was eventually terminated.
- By contrast, manufacturers of PCs and other systems that use microprocessors do not attempt to control the suppliers of these key subsystems, and commercial aircraft manufacturers do not try to control the suppliers of engines or electronic systems. In this way they ensure that they obtain the best possible products at the keenest prices.

Some relationships are strongly influenced by past practice and custom. Whereas the airlines, with few exceptions, have bought what the aircraft manufacturers have provided, the military have a long tradition of specifying their needs and becoming involved in much of the development and production activity*. Military systems are also well known for being expensive, late and relatively unreliable. By contrast, new civil aircraft usually are developed within the promised timescales, do not greatly overrun cost targets and are remarkably reliable.

An important feature of purchaser-supplier relationships when the purchaser specifies the requirements is the process of inviting potential suppliers to bid for the work, and the subsequent task of evaluation and selection. To enable potential bidders to submit proposals it is considered necessary to specify the requirements in detail. If the requirements are not detailed it will not be as easy to

---

* National railways and electric power utilities (until fairly recently, before most were privatized) used to adopt similar policies. Partly this stemmed from the fact that the supply side of these industries used to be owned and controlled by the same people who used their products, i.e. governments. There used to be only limited international markets for trains and power stations, but this has changed. For example, within the European Community major requirements such as these must be advertised in the EC Digest and any competent supplier may bid.

compare bids. The potential suppliers must then submit very detailed proposals, with costs and other details such as management plans, support plans and often many others, resulting in very large amounts of documentation. Bid preparation is nearly always a frenetic task, since there is always limited time and an absolute deadline.   Bid preparation is also a costly task, with no guarantee of obtaining the business. The people who perform the evaluation and selection must work only from the bid documents submitted and must use judgment criteria and weightings decided beforehand. The whole process can be extremely expensive for both sides, and adds considerable time to the project without any forward engineering progress.

When the purchaser is closely involved in project development, he usually also influences the development of subsystems and components.   For example, a military customer might specify an aircraft, but he will also specify the engine, the avionics and the weapons, as well as related systems such as test equipment. He then finds that he must manage the interfaces between all of these. In order to minimize this task, the main contractor, for example the aircraft manufacturer, is nominated as the "prime" contractor, and is tasked with system management and integration. This is inevitably a difficult task, particularly as the subcontractors will already have been selected by the purchaser and cannot be motivated in the same way as if the prime contractor had freedom to select them and to manage their contributions.   Therefore problems inevitably arise across these interfaces, involving both the prime and subcontractors and the purchaser. Such problems can be difficult and expensive to solve, particularly as military purchasers are seldom prepared to reduce the requirements.

"Involved" purchasers usually specify and monitor standards and methods, such as quality system standards and methods for design documentation. There is an inevitable tendency to increase the extent of specification detail, as every concerned function in the purchaser's organization seeks to ensure that its particular requirements are met. There is always more pressure to add requirements than to reduce or minimize them, so it is not unusual for specifications to include conflicting or out-of-date requirements.

When a supplier works to detailed specifications provided by the purchaser, it is difficult for him to prioritize the requirements. However, prioritization is an essential task of design and development. It is inevitable that some features will be more difficult to achieve than others, and some may even be exceeded by relaxing on others. The project team needs the freedom to study such trade-offs and to make the necessary changes. Some purchasers encourage their suppliers to submit ideas for trade-offs, but even this is not as effective as having the freedom to act without external involvement or even prohibition.

Many purchaser-supplier relationships in which the purchaser is closely involved degenerate into antagonism, marked by mutual distrust and unwillingness to share information. The purchaser's role becomes largely a policing operation, in which he seeks to find out whether the supplier is performing to the contract, and the supplier attempts to hide evidence of non-compliance and to report only successes. Of course engineering problems can seldom be hidden for long, so when they do emerge the purchaser's suspicion is increased. In a situation as unconstructive as this, much management and engineering effort is wasted on investigation and argument. Project politics become more visible than actual engineering, so that the best engineers might not find the career satisfaction they desire and deserve. The best engineers do not find this kind of regime satisfying, and therefore tend not to join or do not stay for long. The culture of antagonism therefore develops in the organizations and becomes ingrained in expectations, attitudes and procedures. The effects are apparent in the purchasing organizations, in the suppliers and far down the supply chain. Breaking out of the mould is very difficult and is seldom achieved[*].

The clear lesson that we can learn from these experiences is that purchasers of engineering systems should allow their suppliers the maximum freedom of action. If specifications must be written, they should state the minimum, essential requirements (including costs and timescales). Desirable requirements should be stated and their priority indicated. The supplier should be allowed to manage the entire project, with the absolute minimum of interference from the purchaser. On the other hand, the contract must be structured to provide the greatest possible incentive for successful development and production. This can be provided by making the supplier responsible for the costs of failure to meet the essential requirements and of failures in use, by reliance on normal contract law and warranties.

---

[*] The story of recent UK defence equipment projects seems to indicate that, despite initiatives to improve the process ("smart procurement"), cost and timescale overruns and the related problems are still endemic. Will they ever learn?

However, it is better and more effective to apply positive motivators, such as incentive payments and prospects of further business.

Suppliers of major projects should not be selected on the basis of the conventional procedure of preparing specifications, requesting proposals, evaluating these, selection and contract award. It is far more effective for the purchaser, once he has determined his requirements in outline, to select a supplier whom he knows to be capable and with whom he wants to deal, and to discuss the project. Any specifications needed should then be determined by the supplier, in cooperation with the purchaser. In the early developmental stages the specifications should be kept flexible and should avoid details unless these are essential. When the specification is agreed the contract should be negotiated and placed. The specifications should be finalized and further necessary detail added only when development is complete and production is being planned.

This approach is far faster and more economical than the conventional procedure. It creates a partnership between the purchaser and supplier, instead of the adversarial relationship so prevalent in major projects. However, the purchaser must carefully pre-select the suppliers, and must obtain agreement on the nature of the future working relationship before discussing details. The key elements of the working relationship must be that there will be complete openness regarding costs, methods, progress and problems. There must be agreement that risks will be shared, and risks must be mutually identified and agreed. Payment terms must be flexible to allow for adequate funding to resolve problems that do arise during development. The partnership must be a long-term one, not merely for the duration of the one project: the supplier must have the confidence of long-term business so that he can balance the risks and benefits over a range of work, and can steadily improve his performance with experience. Contracts must cover all phases of the project, including support. (Some purchasing organizations, mainly government departments, actually place contracts for project design and development, then open a new bidding process for production. Such a policy is quite against the principles described here, and is quite ludicrous in the context of complex engineering products). The purchaser must demand that the supplier continuously improves his performance, in all respects, and he must help him to achieve this by mutually agreed training and sharing of information.

In certain cases it might be appropriate to retain more than one supplier for a particular type of product, particularly when it is desirable and cost effective to spread the work. Sharing the work between suppliers should be considered only if they will all be given sufficient work to keep them continuously engaged, without long periods between projects. The different suppliers must be selected and treated in the same way, and the way that work is shared must be agreed. It is a good idea to introduce an element of competition when multiple suppliers are used, related to performance and rewarded by increased share of the business or allowed profit. However, such competition must not be allowed to impair the business of the other suppliers, and there must be agreement for sharing of information on methods used and problems faced and solved. Unlike conventional contracting arrangements, once suppliers are selected they must no longer strive to win business from the joint purchaser at the other's expense except by improvements in quality and service.

Involvement by the purchaser in engineering development is not without potential benefit. In some situations it is inevitable. When the purchaser is the only customer, or for overriding economic reasons must finance the project, or, as in the case of weapon systems there is good reason for stipulating requirements for subsystems, then purchaser involvement is necessary and sensible. The purchaser can help the supplier to optimize the design, both conceptually and in its details. He can be a source of advice on aspects such as trade-offs and support. However, very careful management is required to ensure that the relationship remains helpful and positive. Inevitably there will be disagreements, particularly on details, so clear rules must be laid down on how these will be handled.

Involvement can be made to be beneficial if it is perceived by the supplier to be helpful to him in meeting the purchaser's requirements, and therefore in improving the supplier's prospects of succeeding on the contract. Involvement should therefore be positive, and the people concerned must have the experience and skills necessary for working with the development team.

The same principles apply to the relationships with subcontractors and lower level suppliers. Whether the people concerned are working for the higher-level supplier or for the system purchaser, their involvement must be positive and based on skill and experience. At all levels there must be an attitude of partnership, trust and mutual support.

This approach to managing engineering projects can work effectively only when the managers involved on both sides fully understand and operate the philosophy of industrial partnership. The approach demands high standards of integrity and long-term commitment by the chief executives concerned, and must be supported by careful selection of managers, by training and by supervision. The approach might seem idealistic to managers accustomed to the conventional "scientific" way of doing business, but it is also realistic. The cases quoted earlier illustrate this, and the most advanced and competitive industries, such as airlines, aircraft manufacturers, car manufacturers and electronic equipment companies apply the philosophy, which is totally consistent with Drucker's teaching. An excellent discussion of the application of the philosophy is given in the book by J.P. Womack, D.T. Jones, and D. Roos, *The Machine that Changed the World* [12].

EXTERNAL CONSTRAINTS

Some projects must be managed in ways that are determined by forces outside the project manager's control. This situation often arises in public sector contracting, for which government policy and legislation impose rules and constraints. Foremost among these, in the context of the conclusions of this chapter, is the statutory necessity to comply with complex selection procedures to ensure that every project is open to all potential bidders. (The European Commission has generated strict rules for European public sector purchasing). This prevents the formation of long-term partnerships as described earlier, and forces purchasers and suppliers into bureaucratic, potentially adversarial relationships. Part of the justification for such policies is that they can reduce the chances of dishonest dealing, and public sector contracting must be seen to be "clean". The other justification is that competitive bidding is considered to increase efficiency and to reduce contract costs.

It is true that long-term, non-competitive contract relationships can lead to inefficiency, and even to fraud. However, this is not an inevitable consequence, and it is not beyond comprehension how safeguards could be provided. Of course the most important safeguard is the competence and integrity of the managers concerned. Any other safeguards imply that the managers are not trustworthy. Therefore it is far more effective to apply thinking and effort to selecting and training the managers of projects than to the erection of rules to deter and detect the very few who might transgress.

The disadvantages of competitive tendering have been described earlier. Rules that impose this approach are usually the consequence of past poor performance and cost overruns on non-competitive projects. However, the problems experienced have not been inherent in the method. Rather, they have been the result of poor management, on both sides, including inadequate appreciation of risks and then acrimony when the inevitable problems arise. Competitive tendering does not prevent these problems. The approach encourages the potential suppliers to play down the risks (in consequence, "risk management plans" must now be prepared as part of the proposal on some projects). Projects are awarded on the basis of what is written in the proposal, including the price. Once the contract is awarded and work begins, the supplier's motivation is limited to that inherent in that project, with all of the disadvantages described earlier.

Accepting that some projects are constrained in these ways, and that governments are unlikely to change the rules in the near future, engineering project managers in purchasing and supplying organizations can still work in accordance with many of the principles described earlier. They can ensure that requests for proposals emphasize the need for an open, partnership approach during the project. They can make the competence and record of the bidders part of the selection criteria, they can ensure that the contracts include appropriate motivation for success on essential criteria, and that specifications leave the suppliers freedom to apply the skills for which they are being employed. They can set the limitations and rules for involvement by the purchaser in design and development. Finally, they can manage the purchaser's side of the project effectively.

MONEY

Cost Estimating

Estimating the costs of engineering projects can be difficult and uncertain. The uncertainty is due mainly to the extent of engineering risk involved. Design, development and production of a variant of an existing system entails much less risk than for a new system that is based on unproven technology, and the margin for error in high-risk developments can be very much increased. In addition to the effects of engineering risks, further uncertainty often arises due to the fact that there is nearly always pressure to generate optimistic cost estimates, particularly in competitive bidding situations, and this is coupled with the natural tendency of most people to be optimistic about the outcomes of endeavours on which they work. Cost estimates must often be prepared early, when risks are not fully understood and detailed planning is not possible. Everyone is familiar with the fact that major engineering projects in fields such as defence, transport, applied science and civil engineering often overrun their estimates and budgets by large amounts.* Methods have been developed to assist in the compilation of project cost estimates. Computerized and manual cost estimation methods, based on "cost estimating relationships" (CERs), are available. CERs are mathematical models that relate expected costs to factors such as technology, complexity, risk, and experience. Costs can also be estimated and managed by using an extension of the Programme Evaluation and Review Technique (PERT) described earlier, in which cost estimates are determined for all of the separate tasks that make up the project. This is called PERT/COST. More recently, life cycle cost (LCC) software, also called product lifecycle management (PLM)** software, has been created. Modern business systems software, as discussed in the previous chapter, includes these capabilities.

Formalized methods for project cost estimating can be misleading, for two main reasons. Firstly, they encourage the notion that cost estimation is a specialist task, separate from the actual management of the project. Secondly, the methods cannot predict the nature and extent of risks. They cannot serve as crystal balls. Setting up more "sophisticated" methods and specialist cost estimating and management groups can add to costs without producing better forecasts or control. It is typical for contingencies to be built in to estimates to cover uncertainties, but these are often merely gross mark-ups of the overall cost estimate. Generating precise cost estimates and then adding arbitrary contingencies obviously reduces the credibility of the exercise and of the outcome.

There is no easy solution to the problem of project cost estimation. Since the actual cost of design and development is driven primarily by the number of people involved and the project duration, the project manager should be the person best placed to forecast and influence it. However, it is essential that the estimate is based upon the most objective criteria possible, so other people should also be involved, particularly when risks are high. The problem must be dealt with by adopting an integrated approach, with the key engineering and other managers working together to determine the resources and time

---

* It is interesting that the costs of well-publicized construction projects, where the uncertainties and risks are far lower, often overrun by orders of magnitude. This probably has more to do with management failings than technology.

** PLM does much more than cost analysis. Proprietary programs include aspects such as design interface management, linked to CAE.

needed and, most importantly, to identify the extent of risks involved and how these might influence their estimates. A crucial aspect at this stage is a clear and complete understanding of the requirements to be met, as detailed in specifications, contracts and any other relevant documents, support, standards, etc. Non-engineering staff can provide essential support by advising on aspects such as availability and cost of finance, labour costs and overheads.

The engineering project manager should take final responsibility for the estimate. If financial people take the lead the engineering managers will be less likely to understand the basis for the estimate and to be committed to achieving it. However, this approach requires that engineering managers must be properly instructed and experienced in cost estimating and management, for the types of products involved. They must also be given the responsibility for all decisions that will influence total costs, such as design solutions, test programmes, selection and training of project staff and use of resources. Without this authority they will not be able to control the most important costs for which they bear responsibility.

The costs of manufacturing engineering products include fixed costs such as space, machinery, test equipment, etc., variable costs of materials, processes and people, and other costs such as failures and repairs. These costs must also be estimated in advance. Manufacturing costs are usually easier to predict with reasonable accuracy than are design and development costs, particularly if the product is being designed to meet a particular manufacturing cost target. If the product is to be made in quantity the effects of early problems and learning can be overcome, and cost estimating is therefore more accurate for large production runs. There can still be areas of uncertainty, for example if new processes are to be used or if they are likely to create high proportions of defective products. The same principles, skills and experience must be applied to production cost estimating as for design and development, but of course the engineers who will plan and manage the production processes must take the lead.

Minimising Total Costs

A typical engineering project incurs high initial costs of design, development and production, before sales begin to generate revenue. However, sales also generate costs: there might be costs of maintenance and other support and failures in service must be paid for in terms of warranty, replacements and possibly lost sales due to loss of reputation. Hopefully the costs will be repaid and there will be a positive return on the investments and continuing sales and profits. This all happens over quite long and overlapping periods of time, typically one to three years before the first sale, one to two years into warranty and for several years thereafter.

We can minimize the up-front costs of design, development and production. For example, it might be decided to reduce the effort on training, reviewing designs for correctness and testing, all very common expedients. However, the impact of these short term economies on subsequent costs must be considered.

The objective of training, design analysis and test, as described in later chapters, is to create excellent designs that will cost less to produce and support, and that will therefore generate more sales. Therefore, any reductions in these activities might result in higher costs downstream and higher total life cycle costs (LCC). Figure 6.1 illustrates typical hypothetical scenarios. However, it does not show the further likely effects of up-front effort on reducing the time needed for development testing, where large savings are often generated because fewer problems and delays are encountered, so time to market is reduced.

Figure 6.1 Life cycle costs

Whilst it is always quite easy to identify and quantify the short term savings that can be made, estimating the downstream effects is very uncertain. Also, the downstream effects are felt in the future, mostly some years later, beyond the scope of many business financial forecasts. Therefore, it can be difficult to generate and maintain the top-level management support that is necessary. However, managing engineering projects is a long term business, so managers at all levels must understand the total life implications of decisions. The methods described in the subsequent chapters are all proven to minimize total life costs and to maximize the chances for success. We will return to this theme in more detail later.

CONCLUSIONS

Managing engineering projects is one of the most challenging and difficult tasks that people perform. No other tasks present the combination of so much uncertainty, so much dependence on the performance of other people and organizations, and such a range of disciplines. Engineering projects are also subject to severe constraints of time and resources, and must be managed in the context of rapidly developing methods and technologies. Project leaders need to have the knowledge and skills to make the diverse and changing team perform effectively, and to manage the external aspects, such as contracts, funding, reporting, and sometimes politics. All engineering projects combine these features, but obviously it is the large projects that are the most visible, particularly when public money is involved.

The first task of the project leader must be to minimize uncertainty and risk. Time must be provided at the start of the project for thinking and planning. Designs must be analyzed and tested, as described in the following chapters. He must ensure that priority is given to the total project cost over its expected life, and that pressures to make short term savings do not jeopardize this objective.

Next, the project team must comprise as much experience as possible, and training needs must be identified. Even apparently simple engineering tasks take longer and cause problems when performed for the first time. Then the team must be made to play. The temptation to plan and monitor in fine detail must be avoided. A large proportion of engineering development work is in the details of design, test and problem solving, and it is counter-productive to attempt to plan and control every aspect of such fast-moving and unpredictable work.

The project leader must know what is happening, and he must control the game. Most importantly, he must ensure that his team wins. Any activities that detract from success must be put aside. When such activities are nevertheless necessary, for example formal reports on progress or customer audits, they should be assigned to supporting staff, and their interference with the people working on the project should be minimized.

An intangible and unteachable but essential ingredient in any engineering project must be pride. Without pride, and therefore confidence in success, no project can succeed. The work might eventually be completed, but it will probably be late, expensive and sub-optimal. The project leader must instill pride in the team, and he must ensure that problems are dealt with in a way that generates solutions and therefore reinforces pride. The key to this is the devolution of responsibility to the team players (lower level managers, engineers, contractors and suppliers), as described earlier.

Team pride and therefore motivation can be seriously degraded by problems and setbacks, particularly if these are subject to external criticism such as adverse media comment. The quality of management on projects that face such scrutiny must be excellent, and it should include the capacity for effective public and media relations.

Finally, some might protest that there simply are not enough people available with the skills and experience necessary for such exemplary performance, making it necessary to impose constraints and controls. The answer to this can only be another question: Why are such managers not available? Every organization, public or private, has the responsibility to select and develop its engineering project leaders in accordance with the principles described in the preceding chapters.

7

DESIGN

Engineering design is the creative process of identifying what needs to be made and how it is to be made. This encompasses a wide range of activities, from the initial concept to detailed consideration of costs, materials, parts, parameters, dimensions, production processes, use and maintenance. There can be other considerations, such as safety, appearance and compliance with standards.

Development covers all the activities needed to show that the design will achieve the aims and to show up any features of the designs that require to be changed. Development work consists primarily of analysis and testing, recording and analyzing the results, and any necessary refinement or improvement of the designs. Design and development are therefore closely linked sets of activities, which occur iteratively and in parallel from the time that any aspect of the new product begins to be analyzed and tested. It follows that the people involved in design and development must be organized into a single, integrated team.

THE MARKET FOR ENGINEERING PRODUCTS

The product to be designed must be determined by the market. For some products the purchaser or purchasers specify what is needed, in varying degrees of detail, and the designer's first task is then performed to this extent by the purchaser. For example, military systems are usually specified in detail by the government departments that fund the programmes, and a car manufacturer might prepare a detailed specification for an engine fuel control system. At the other extreme, purchasers of products such as dishwashers and hi-fi systems have no direct links with the designers, who must therefore anticipate and influence the market for the products.

The products of engineering design and development serve two basic market areas. Some products are specified and bought primarily by other engineers, or by other professional buyers. Electronic components, rivets, hydraulic actuators and plastic mouldings are examples of the great range of products that engineers use to create higher level products such as aircraft, lift trucks and electronic instruments. These are bought by engineers or by other professional users. For these types of product the designers can usually anticipate the market to a large extent. Suppliers and purchasers can communicate easily, and both usually have a good understanding of the requirements, possibilities and limitations of the designs. Engineers in the purchasing organizations or in the suppliers' service functions often work closely with the design and development engineers, and this is always beneficial. The distinguishing feature of this type of purchaser-supplier relationship is the mutual professional understanding. In turn, this often results in long-term preferences, possibly over a range of products, and price may not be a dominant consideration that can greatly influence the market.

The other type of product is created by engineers for customers who are mostly not professional in terms of their buying decisions. Such products range from cars to cigarette lighters, and from TV sets to toys. The customers do not specify their requirements or work with the designers. Their choices are determined to a large extent by factors such as style, fashion, marketing and price, any of which can greatly influence the market, which therefore becomes much more difficult to forecast. There is little or no loyalty to particular suppliers, and relationships are therefore limited to the initial buying decisions and to any subsequent contact such as for repair.

There are of course some products, such as personal computers, utility vehicles and power tools which serve both types of market, but the distinction is nearly always quite marked, and where there is overlap the non-professional factors are less significant.

It is essential that designers appreciate the type of market at which their new products will be aimed, and that they design accordingly. This is usually not too difficult for professional products, but many engineers have difficulty in relating design to the non-professional market. There are many examples of new products that have incorporated engineering advances but which have been commercial failures because the innovations did not appeal to the intended market, or were not adequately marketed.

I am writing this (1993) on an ingenious modern product called the AgendA. It is a miniature electronic organizer with a built-in word processor. The feature that makes it unique is that, in addition to the conventional layout of small keys, it includes five keys laid out so that a finger rests naturally on each. By pressing simultaneous combinations of these (plus two others close to the thumb position), every ASCII character and common word processing function can be generated. It is far easier to learn to use, much more compact than a conventional QWERTY keyboard, and just as fast after very little practice. The very compact layout makes the whole machine very small and therefore easy to carry and to use when traveling. It is also much cheaper than a laptop or notebook computer. The breakthrough is in the keyboard design, coupled with the necessary software to create characters and functions from the multiple key presses. The only reason that competitive products use the conventional layout is because the original mechanical keyboards could not accommodate the principle of multiple key presses, but this is of course not a problem with modem electronics. However, the AgendA product is little known, and created only a very small niche market. By contrast, compact electronic organizers which use conventional keypads dominated this market, even though they had no word-processing capability, and data input was awkward because of the very small keyboards. The AgendA is an example of an excellent engineering design, incorporating a significant new feature, which did not succeed because of the fickle nature of the market for this kind of product. More recently, electronic organizers with handwriting inputs have become available, but it is interesting that the AgendA's multi-key character entry concept has never been copied or extended, and the product disappeared[*].

On the other hand, the Sony Walkman was a classic example of a new product which involved little real engineering innovation, but which nevertheless created a vast new market.

There are many other examples of failures and successes of engineering designs. The "revolutionary" British Sinclair C5 electric vehicle failed totally because the driver sat so low that he felt vulnerable in traffic, and comparable engine-driven vehicles were faster and more versatile. The early Hewlett-Packard touch-screen computer was unsuccessful because users had become accustomed to using the "mouse" pointing device, which was faster and more ergonomically appealing. Successes include the IBM personal computer, which set the standard for PCs even though it did not outperform its competition technically, but created a huge dedicated software industry, and the Texas Instruments "Speak and Spell" toys, which exploited early speech-generating integrated circuits in away not originally planned by their developers. More recent notable successes include Dyson's vortex vacuum cleaners, mobile phone

---

[*]I still use mine!

systems and the Apple i-Pod. The Motorola Iridium satellite communications system was a failure, though, because very few potential subscribers were prepared to pay the high charges that would enable them to have global coverage, when much cheaper land-based networks like GSM could cover the great majority of the populated areas of the world.

The lesson of the successes and failures is that new product designers must create what people will buy (or can be persuaded to buy), and not just what the designers (or the companies that employ them) find intriguing or challenging. The spurious challenge of misdirected innovation can be appealing to product managers and designers, but can lead to expensive failures.

Initial design, to avoid failure and maximize success, must be based upon what the market wants or can be made to want. Determining what the market will want is principally a deductive process, even when innovation is involved (contrast the AgendA with the Speak and Spell). The deductive process must take account of every factor that will influence the market: appearance, fashion, price, quality, performance, competition, safety, and many others depending on the product. Any of these factors, or combinations of them, can make or break the product, but their effects are often uncertain and must be predicted, often years in advance. Initial design must therefore be managed as a totally integrated activity, which ensures that every aspect is considered and optimized in relation to customer perceptions of the product.

Product innovation must be linked to the marketing strategy, which must emphasize the features that differentiate the product from the competition. The AgendA provides a good example of failure to link the primary innovation to the advertising campaign: the AgendA was advertised as an electronic organizer, with no emphasis given to its major advantage of rapid, easy data input using the "microwriting" keys. Marketing and advertising will not make up for conceptual mistakes, such as the Sinclair C5, but it is essential to generate awareness of the features that will sell the product.

PRODUCT STRATEGY

Design and development must be conducted within a strategic framework, which must determine what kinds of products will be created. The strategy enables resources, training and priorities to be determined, and provides the basis for longer-term planning. For established organizations the strategy would normally be to provide the kinds of products made in the past, since these represent the knowledge of the organization. It is also to a large extent what the markets will expect. For example, a manufacturer of cars or of hi-fi equipment would normally have such a strategy. This basic strategy should be further refined, however, to provide particular direction and motivation A car company might decide to develop a new family of sports cars or off-road vehicles, and a hi-fi company might decide to concentrate on the top end of the market. A strategy can be expressed powerfully in relation to the competition: when Komatsu decided to compete strongly with Caterpillar in the world market for earthmoving equipment they expressed their strategy, to all of their employees, as "beat Caterpillar".

Product strategy must be based on the engineering strengths available, in relation to the strengths of the competition and of suppliers. These strengths are primarily knowledge, particularly if patents are held, capability, and position in the market. The strengths must be considered not only in relation to what exists now, but also to the situation as expected and intended in the future.

DESIGN OPTIMISATION AND INNOVATION

Benchmarking

"Benchmarking" is a term used for the process of identifying the features of a product that will appeal to customers, and then comparing the features with the equivalent features of competitive products. Each feature is rated for importance and in relation to how well the products achieve them. In this way the new product and the competition can be compared systematically, and priorities can be assigned to the achievement of the most critical features. The benchmarking process is carried out by a team that combines knowledge of the market requirements, the technologies available, competitor capabilities and any other pertinent information. Sometimes customers or potential customers are used to assist the process, and competitive products should be obtained and tested when appropriate.

For new product designs the process of differentiation from the competition should be taken further than this, to identify the features that should be developed to distinguish the product, not merely by comparison with equivalent competitor features but which are different in kind. Such features can be more effective in generating sales than improvements to existing expected features. They can also provide further strategic advantage, particularly if they can be patented. For example, the Shimano indexing bicycle gear system introduced a new feature to the traditional Derailleur system, and created a huge market.

Benchmarking is entirely a deductive process, but innovations are mainly inductive. The highly structured approach applied to benchmarking can stifle innovation, so the product design process must also be made to stimulate innovation. Innovative, inductive thinking can be encouraged during the early design stages by making sure that questions regarding product differentiation are stated from the point of view of the customer, and not merely in relation to product features. For example, people at Shimano might have asked: "How can we reduce the variation between gear change positions?" Instead, they addressed the question: "How can we make gear changing easier?"

Benchmarking seldom involves new technology or methods, but innovation might. Also, innovation can introduce risks in relation to cost, development effort, reliability and product acceptability. Therefore all innovations should be subjected to the careful, deductive appraisal inherent in the benchmarking process, and in the methods described later.

The benchmarking approach can also be used in relation to processes or other capabilities, to help to identify where competitive improvements can be made, for example in process yields, research effort, staff morale, or any other aspects that differentiate competing organizations.

Quality Function Deployment

Quality Function Deployment (QFD) is a technique to assist in the systematic consideration of all aspects of the design of a product, including production processes, from the point of view of the expected customer, and the methods, responsibilities, and controls that will be necessary.

The QFD process uses a chart, as shown for a design for an electro-hydraulic motion system in Figure 7.1. The product requirements are listed and rated for importance to the customer: they are sometimes referred to as the "voice of the customer" (VOC). The product features that will affect the requirements are tabulated. Each feature is rated in relation to its contribution to each requirement, and a total rating for each feature is derived by adding the products of the requirement importance values and the feature effect rating. Other aspects can also be added, such as options to be considered, responsibilities, etc.

	Requirements	Units	Importance	1 Selector	2 PWM controller	3 Solenoid valve	4 Actuator	5 Position sensor	6 Interlock	7 Pressure switch	8 Changeover valve	Comments	BENCHMARKS
	Interactions ✿			2	1,3	2	3	2,4	2	8	7		
1	Response rate	15Hz	5	0	4	4	4	2	0	0	0	Analyse, test	4
2	Max. thrust	200N	5	0	0	0	5	0	0	0	0		2
3	Accuracy	0.5mm	4	2	2	4	2	5	0	0	0	Analyse, test	3
4	Max overshoot	10mm	2	0	0	3	2	2	5	0	0	Function test	2
5	Fail safe		5	2	4	3	4	2	5	5	5	Analyse, test	3
6	Hyd supply changeover	150 bar	4	0	0	0	0	0	0	5	5	Hydraulics test	1
7	Max. temp	45deg C	4	2	3	2	2	3	0	2	2	Accelerated test	1
8	Min. temp	-20deg C	4	2	4	2	2	4	0	2	2	Environ-Mental test	1
9	Max. weight	12Kg	3	3	1	3	4	2	1	2	2		2
10	Reliability	Very high	5	3	3	4	3	4	2	3	3	Accelerated test	3
11	Durability	20 years	4	4	2	4	4	4	2	3	3	Accelerated test	3
12	Max. production cost	$400	4	3	2	4	4	2	1	2	3		4
			SCORE	86	110	134	152	122	70	102	106		
			NOTES	1	1								

Notes: 1. Consider/analyse continuous servo control (accuracy, cost) in place of PWM.

Figure 7.1 QFD of motion system

The interactions columns show which requirements and features influence one another. For example, in the example the pressure switch and the changeover valve strongly influence one another, since they both contribute strongly to the requirements for fail safe and hydraulic supply changeover (shown highlighted).

Benchmarking, as described earlier, is an essential feature of the QFD method, to consider the extent to which competitive products achieve or are expected to achieve the requirements. A separate benchmark column should be produced for each competitor. The benchmark values are used to assist in assigning importance ratings to the requirements. The new product design should attempt to match or exceed all of the important benchmarked requirements.

The QFD is prepared by a team consisting of all of the people who need to be involved, such as marketing, customer service, design and production. It is essential that everyone who can influence every requirement and feature is involved. The discussion is structured as a "brainstorm", in which all present are invited to contribute, with one person maintaining the record of contributions. Only when no further ideas are being contributed is discussion on relative importance allowed. This ensures that contributions are constructive and that no one is reluctant to contribute.

When the top-level QFD has been performed, each feature is then subjected to a separate lower level analysis to identify the detailed features relevant to ensuring that it meets its separate requirements. This "part-level" QFD identifies specific materials, processes, etc. appropriate to the parts of the product. For example, the analysis of the actuator will consider available types, materials, dimensions, costs, etc. The analysis continues downwards until the final "process-level" QFD considers detailed manufacturing processes and their controls, such as statistical process control limits, inspection, etc. In this way every top-level, customer-perceived requirement is "deployed" down to the detailed specifications, processes and controls that will be used in all stages of design, development and manufacture.

QFD can be a very effective method to ensure that all product requirements are systematically and rationally considered and managed. It forces an integrated approach to product design, development and production, and can greatly reduce the likelihood of unpleasant surprises in later and therefore more expensive project stages. Time spent on the analysis might seem initially to be non-productive, but such early, careful study will always pay dividends as the project proceeds.

TRIZ

TRIZ (a Russian acronym for "Theory of Inventive Problem Solving") is a systematic approach for generating innovations. It was developed by Genrich Altshuller. Check it out.

"Six Thinking Hats"

Edward de Bono created the "six thinking hats" approach to creative thinking. The "thinking hats" represent different aspects of personality and opinion, such as objectivity, intuition, pessimism, rationality and emotion. Check it out.

Pugh Matrix

The Pugh matrix, or "Criteria Based Matrix", is a technique to help determine which items or potential solutions are more important or 'better' than others. The method facilitates a disciplined, team-based process for concept generation and selection. In the matrix options are assigned scores relative to criteria. Several concepts are evaluated according to their strengths and weaknesses against a reference concept called the datum (base concept). The datum is the best current concept at each iteration of the matrix. The method is similar to QFD, and in some ways it is complementary.

DIVERSITY AND AGILITY

Product concepts developed by innovation and deductive methods should be extended to cover as wide a potential market as possible, and to be adaptable to changes in market requirements, or even to improvements if the original concept needs to be changed. This flexibility should be retained in the design for as long as possible. The aspects of the design that might require to be changed or varied must be identified, and scope must be left for introducing the changes or variations at minimum cost and effort.

In this way products can be launched with a diversity of features, ensuring that they appeal to a wider section of the market, and changes can be more easily made in order to respond to market demand. Design for diversity and agility can greatly increase market share and the time that a product line remains competitive.

DESIGN OF LOGIC-BASED PRODUCTS

Most modern engineering products include digital electronics and software. The design of logic-based systems and subsystems therefore requires the simultaneous design, development and integration of these elements. There are aspects of digital electronic and software engineering that are new and unique in relation to much of engineering tradition. Therefore the ways in which they are managed as part of the overall design and development effort must be considered.

Digital hardware and software logic designs share the property of being either correct or incorrect, and this property does not change from copy to copy or with time (for most practical purposes). Other engineering hardware suffers from variability in production and from time-dependent changes, such as wear, corrosion, etc., as will be discussed in Chapter 10. (Digital electronic hardware does, of course, suffer from production variability, but usually to a much lower extent than most other products as far as the function is concerned. Digital circuits, once built and tested, rarely change or fail). Therefore, the main objectives of logic design are to create correct specifications and to interpret them correctly. However, correctness and the converse, the existence of errors, are not usually easy to observe. Logic systems lack the "perceivability" of most other designs, whose correctness can usually be checked by examination of the hardware or drawings or by checking calculations. Logic specifications and designs (programs) must be managed with care and discipline to minimize the creation of errors.

The design and development of digital electronic systems is a process that is strongly supported by automation, owing to the relative ease of computer simulation of logical systems. Electronic design automation (EDA) software can be used to synthesize the system and test it, as well as automating the design of the placement and connections of components and generating the software for automatic assembly and test. The design is also automatically fully documented, by drawings, component lists, connections and test instructions. In principle, the designer need know only the input and output logic and which component building blocks are required, since the EDA programs contain the functional models of all of the components to be found in the catalogues. The design of such circuits, particularly when part or all of the functions are implemented on application-specific integrated circuits (ASICs), rather than by using separate standard circuits, is increasingly performed by specialist groups, often working as external suppliers to the project. It can therefore be difficult to integrate this work into the rest of the design.

Software design and development is not automated to anything like the same extent. The logic within software is much more complex than the basic functions available in digital circuit elements, and there are no standard software "subsystems" equivalent to, say, an analogue-to-digital converter or a signal processing circuit. Software design must therefore proceed from the system logic requirements down to the individual code instructions. The only documentation that is automatically generated is the actual code, and this is almost impossible to understand and review without additional information such as detailed user notes. Whilst each copy of a program will be identical to all others, no two programs can be the same. Therefore the potential for automating testing is very limited. Such software testers that have been developed can detect various logical and syntactical errors, but they cannot test whether a particular program actually works over its operating range of inputs, outputs and conditions. Therefore software testing has to be planned and performed with a high degree of thoroughness and disciplines. Software testing is discussed in more detail in Chapter 8.

The hardware and software meet in the system processor and its memory. Processors and memory are nearly always standard components, and the same processor and memory can handle any program that is written for it. Software-hardware interfaces also exist at points such as input devices, sensors and displays. Other important interfaces, particularly with modern systems using faster, more densely packaged electronics, are timing and electromagnetic compatibility (EMC). The software designers must therefore understand the possibilities and limitations involved, and the hardware designers must understand the software implications.

The design team should, therefore, as far as is practicable, include logic (hardware and software) design and development as activities integral to the project, not as specialist tasks to be performed by people outside the team. Optimizing the balance between functions to be performed by the main technologies (electronic, both analogue and digital, non-electronic and software) is an essential aspect of overall system design, and this can be achieved by an integrated team in which opportunities and problems can be readily discussed and ideas can be quickly tested.

DESIGN FOR PRODUCTION AND MAINTENANCE

Selection and optimization of the processes to be used for production must be an integral part of the design and development activity. The design team must know the processes that are necessary, the methods available for implementation and their advantages and limitations. The advantages and limitations must be considered in relation to the quantity to be produced, cost objectives, requirements for repeatability, diversity, accuracy and measurement, and other special aspects. The product must be designed for the processes that will be used, and the processes must be designed and developed concurrently with the design and development of the product.

The importance of production aspects of design and development obviously depends upon the extent to which production will affect costs and markets. For items that will be made in large quantities for competitive markets, such as power tools, TV sets, cars, mobile phones, and the parts that go into them, the costs of production will critically influence competitiveness, so design and development for production of such products is therefore as important as the achievement of performance objectives. However, even for few-of-a-kind or one-off products, ease of manufacture can greatly reduce costs, both during development work and later. Therefore every designer must be concerned with how his design will be made, assembled and tested. For items that will be maintained and repaired, there is a strong connection between ease of manufacture and ease, and therefore cost, of maintenance and repair, so these aspects should be considered together when appropriate. As far as is practicable, the designers should be available to help to solve problems that might arise during later stages, even if they might have moved on to other work.

Production processes can be performed either by people or by machines. People can perform an infinite variety of production tasks, but they are limited in terms of power, speed, ability to perform repetitive tasks accurately, and sometimes by environmental factors such as temperature, noise or radiation. People are usually more adept than machines at complex assembly tasks, some measurement tasks, and when versatility or other particular human senses and skills such as vision and judgment are required.

Many processes are performed by machines controlled by people, the machines providing capabilities that humans cannot achieve, such as the ability to shape metals quickly and accurately by cutting, grinding, or pressing. Other processes are performed completely by machines, the only human involvement being the initial set-up and monitoring. Body-welding robots, multi-function machining centres, electronic component placement and automatic electronic testing are examples of such automated processes.

Flexibility is an important feature of production systems, if there is any requirement to produce different items, products with different features or if improvements have to be made, either to the product or the process. Processes operated by people are inherently flexible. However, processes performed by machines and automated processes are often specialized, and process design and development must ensure that the right balance is struck between flexibility and specialization. A classic example is given in the book *The Machine that Changed the World* [8] of the development of car body panel presses by Toyota to enable rapid change of dies so that different panels could be produced from each press, depending on the exact requirements of particular models and variants. Flexibility can greatly increase the cost justification for automation, as well as enabling much more rapid responses to changes and product diversity. Of course software-controlled systems have a large degree of flexibility built in. Numerically controlled machining centres can make an infinite variety of shapes within their limitations, and electronic component placement machines can be used to assemble any circuit, again within the limitations of the system.

Design For Machine Production

The most important factor in designing for production by any machine is the capability of the process or processes involved. Process capability includes production rate, accuracy, cost and any other features necessary to the product. If the process is a standard one, for example a grinding operation or electronic component soldering, then no process design is involved: the designer needs only to understand all aspects of the process capability and how they can affect the product.

If the process is not standard, or if standard processes cannot provide the capability needed, then special processes must be designed and developed as part of the overall design and development activity.

Design For Manual Assembly

People have far greater versatility than machines, so tasks designed for them can exploit their wide capabilities and flexibility. On the other hand, people as process operators are expensive, and they are much more variable than machines. Design for human tasks must therefore minimize the time taken and the possibilities for error. Careful attention must be paid to "mistake-proofing", by designing the product and the processes so that operations can be performed only the one correct way, or so that incorrect operations will always be detected and prevented from being passed to the next production stage. In Japan mistake-proofing is called *"Poka Yoke"*. "Murphy's Law" states that if a task can conceivably be performed more than one way, and only one way is the right way, then someone, some time will do it the wrong way. Obviating "Murphy's" requires close involvement of production people in the design of the product and of the manufacturing processes.

DESIGN REVIEW METHODS

It is unlikely that the design of a typical engineering product will be right in every detail the first time. Effective management and training will greatly reduce the chances of error and oversight, but so many aspects must be considered, especially for designs involving multiple disciplines and interfaces, that errors and the need for changes will be inevitable. The problem is of course compounded when new technologies or methods are being used, and if the design must be produced in quantity. For most modern products the time from concept to sales must be minimized in order to be competitive, and there is little or no time for redesign if problems are discovered later in development. Finding and correcting design errors at later project stages becomes progressively much more expensive (the "times ten rule" mentioned in Chapter 8 applies). Therefore it is important to review designs to ensure that any errors and oversights are detected and corrected early.

There are several formalized approaches to reviewing designs. These provide methods to check designs in relation to aspects such as reliability, safety, stress resistance, corrosion protection, ease of manufacture and test, and any other features that are important. Such disciplined analyses are essential for difficult projects, regardless of the team organization or skills. The analyses must follow a disciplined sequence and must cover all parts of the product design. For example, all stresses applied (electrical, thermal, mechanical, etc.) to each component should be identified and evaluated in relation to strength, durability and reliability. All electronic designs should be analyzed for testability, electromagnetic compatibility and parameter variations. All software should be reviewed to check detailed compliance with the specification and absence of errors. Failure modes, effects and criticality analyses (FMECA) should be performed on all designs and processes to identify likely failures, their causes and effects, and possible action to prevent them or mitigate their effects.

These analyses must be scheduled as mandatory milestones, to be performed and documented before proceeding to subsequent important stages such as release of designs for production, and in sufficient time for corrective action to be taken. They take time and occupy resources, but it is rare for well-conducted design review activities such as these not to justify themselves in terms of problems discovered and solved, and in reduction of overall project time and cost. They should be performed by the engineering team, but suitably skilled independent engineers can also contribute or can perform the analyses in close cooperation with the project team.

The separate reviews should be planned as subsidiary tasks to the main project design reviews at which the project leader and other appropriate managers review progress and authorize subsequent stages. Project design reviews must be planned well in advance, and they must be for decision-making, not merely reporting. They must avoid becoming involved in details, which should have been covered earlier by the subsidiary reviews. For important decisions, such as authorization of more funding, action on major problems or release to production, the appropriate managers must be present and must all agree with and support the decisions made.

COMPUTER-AIDED ENGINEERING

Nowadays most engineering design is created with the aid of computers. A range of computer-aided engineering (CAE) software is available (or computer-aided design (CAD), or electronic design automation (EDA) in the context of design of electronic circuits). These range from simple 2-D drawing and circuit design to comprehensive multi-technology systems. The capabilities of CAE allow most design tasks to be completed much more quickly than by manual methods. CAE is essential for the design of very complex modern products such as vehicles, spacecraft and large integrated circuits. In addition to automating tasks such as drawing, routing of electronic connections and performing analyses of stress, vibration, electronic performance and a range of other functions, modem CAE increasingly supports integrated engineering. This is more advanced in relation to electronic design, but CAE software which provides the capability to model functional elements such as springs, actuators and sensors is also available, allowing true multi-technology CAE-based design. Both electronic and mechanical CAE can cater for the interface between design and production, as well as aspects such as variation analysis, electromagnetic compatibility and modeling the behaviour of failed components.

There is also a range of specialized CAE/EDA software, covering aspects such as finite element (FE) based methods for modeling stress/strain behaviour, heat and fluid flow and electromagnetics. Mathematics software includes capabilities for system modeling. There is an increasing trend for the different software systems to overlap and merge.

CAE greatly increases the capability of engineers to design new products rapidly, and to experiment and improve without the need to perform off-line calculations or build and test models. Therefore it can save time and money and help to create better designs. When designs are created using CAE they should be analysed as thoroughly as the CAE system will allow. For example, if the program being used contains the capability for analysing the effects of electronic parameter variation this should be used to evaluate the effects of all important variables and tolerances. CAE should be used in support of failure modes, effects and criticality analysis. In every case the use of CAE for design analysis largely removes the possibility of human error, and it can also provide automatic documentation.

CAE does not solve all problems, and there are limitations in all systems. The component functional models do not always simulate all aspects of reality: for example, most EDA software models for electronic components do not describe the reactions to electrical or thermal overstress or take account of three-dimensional aspects of internal electromagnetic radiation. Therefore the engineers using CAE systems must be aware of the limitations as well as of the capabilities. Full knowledge of capabilities, limitations and new developments requires continuous training, and the amount of training necessary is much more than has traditionally been considered appropriate for design engineers.

Exploiting the full capabilities of CAE is possible only if the engineers using it are thoroughly trained, and if its use is fully integrated into the working methods of the whole design team. This requires top-down direction and support. Ideally, the initial design concept should be generated on the CAE system and, as far as is practicable, all lower level designs and analyses should be performed using the same system. This makes it possible for interfaces to be designed and analyzed concurrently with system and subsystem design. It also improves the accuracy and completeness of system-level analysis and viewing.

Reference 14 includes more detailed descriptions of CAE-based design analysis. The book's homepage (www.pat-oconnor.co.uk/testengineering.htm) includes lists of CAE systems and suppliers.

CONCLUSIONS

This chapter has given a brief overview of the most important aspects of managing engineering design. The books by Don Clausing, *Total Quality Development* [13] and Donald G. Reinertsen, *Managing the Design Factory* [14] provide deeper guidance on managing a product design organization for best results.

8

DEVELOPMENT TESTING

Testing is usually the most expensive and time-consuming part of engineering development programmes. Paradoxically, most testing should in principle be unnecessary, since testing is performed primarily to confirm that the design works as intended, or to show up what features need to be changed in order to make it work. If we could have complete faith in the design process we could greatly reduce the need for testing. There are some products where in fact no testing is carried out: civil engineering structures and buildings are not tested (though many such designs are now analyzed using CAE simulations) partly because of the impracticability of building and testing prototypes, but also because the designs are relatively simple and conservative. However, nearly all engineering designs must be tested. Unfortunately, this is seldom done as effectively as it could be.

The main reason for insufficient or inappropriate testing seems to be that engineers have not developed a consistent philosophy and methodology for this essential activity. Testing is not taught as part of most engineering curricula, and until very recently no books on the wider aspects of the subject had been published. Reference [14]* is the only book that provides an overview of test methods, economics and management You are reading the only book on engineering management (to the best of my knowledge) that discusses the subject. Academics seem to be unaware of its importance, or even sometimes of its existence as a project activity.

Specialist areas are taught, for example fatigue testing to mechanical engineers and digital circuit testing to electronics engineers.   However, a wide range is untaught, particularly multidisciplinary, systems and management aspects. Engineering training tends to emphasise design. Testing (and manufacturing) are topics that attract less attention, and they do not have the "glamour" of research and design. This is reflected in the generally lower esteem,

---

* P.D.T. O'Connor, *Test Engineering* [15].

status and salaries of engineers working in test and manufacturing. In some countries the near-disappearance of technician and apprentice training as routes to recognised engineering qualification has greatly reinforced this unfortunate trend. Engineering industry suffers shortages of talented engineers in these key areas. As a result, designs are often inadequately tested in development and products are inadequately tested in manufacture and maintenance. This creates high costs throughout the product cycle, damages competitiveness, and can lead to hazards.

If the design team possesses all the knowledge and resources (CAE, time, etc.) necessary to create correct designs, and the project leader has faith in this knowledge, then the need for testing can be reduced. Furthermore, what testing is performed will be less likely to show design errors, so the total design effort will be reduced. Knowledge and training are discussed in Chapters 3, 4 and 12, and design facilities in Chapter 7. The point to be stressed here is that the potential improvements in engineering productivity that can in principle be achieved by harnessing the innate ability of people to learn, and then to use their knowledge to reduce the need for product test and redesign is enormous. Nevertheless, despite the most determined attempts to minimise the need for test by team organization, training, analysis and simulation, most engineering product development must involve considerable testing. Design, test, redesign and re-test proceed iteratively and in parallel, at different levels and in different locations.

Therefore development testing must be managed as an integral aspect of the whole product process. Design and test must be closely integrated from the earliest stages, and designers should be active participants in the analysis and testing of their designs. Suppliers' test programmes and methods must also be managed as part of the overall project.

Testing is also an integral part of the manufacturing process, and the methods to be applied must therefore be designed and tested during the preceding phases. Design teams should be aware of the relevant technologies and methods.

Whilst development (and manufacturing) testing is expensive, insufficient or inadequate testing can be far more costly later, often by orders of magnitude. Therefore the test programme must be planned and financed as a long-term investment, not merely as a short-term cost. This can be a difficult concept to sell, particularly as so many organizations are driven by short-term financial measures like end-of-year profits, dividends and stock options. Engineering as well as commercial experience and judgement must be applied to the difficult and uncertain business of test. Managers at all levels and in all contributing functions must appreciate the concept that test is an investment which must be planned, and which can generate very large returns. **Test adds value.**

THE NEED FOR DEVELOPMENT TESTING

Every newly developed engineering project must be tested prior to being put into production or use, to confirm that the design meets the requirements of performance, safety, durability and reliability. In principle, the design should be capable of meeting all the requirements without the need for testing to confirm this. Testing is usually the most expensive part of any engineering project. It requires the manufacture of prototypes, conduct of the tests, often using expensive facilities, and redesign and retest when the need for changes is identified. Therefore the test programme must be carefully planned to generate the maximum value to the project, within the constraints of time and cost.

The need for testing is a direct reflection of the extent of uncertainty in the design. For some new products there is little or no uncertainty, and so testing can be minimal. For example, a purely static item such as a mounting bracket for a stationary component can be designed with confidence that all features that affect performance and durability are understood and taken into account, since the task is simple in engineering terms. If, however, the bracket is to be used to support a component subjected to vibration, and it must also be as light as possible, then uncertainties begin to grow. If the consequences of failure are severe it becomes even more prudent to perform tests to confirm the design calculations and analyses. Additional risks are incurred if the product is to be made in quantity, since the manufacturing processes will introduce variation and therefore further uncertainty.

Of course few engineering products are as simple as a mounting bracket. Even quite simple designs, such as a door actuator or an electronic timing circuit, will contain features that should be confirmed by testing. The amount of testing will obviously depend upon the designers' familiarity with the problems and their knowledge and skill. If the designs are shown by test to be correct then no redesign will be necessary, otherwise changes will have to be made and the changed design retested. Immediately we can see the strong connection between the experience, knowledge and skill of the design teams and the cost of development.

Most engineering products are more complex than the examples given above. In the majority of cases it is not practicable for the design team to possess such knowledge that all problems can be solved and all features optimized without testing, analysing the results and making changes. In fact the process is often iterative over several cycles, especially for difficult or novel designs.   Two aspects in particular add to the need for test: reliability (or durability) and variation.

Reliability and Durability

The reliability and durability of a product depend upon how it is affected during its life by the way it is used and by the environments it must withstand.   They also depend on features of the product that might change with time and thereby affect its performance or strength. Wear of moving parts, fatigue weakening of parts subject to cyclic stress, deterioration of insulation and corrosion are examples of the many processes that can cause an initially good product to fail later in its life. The rate and extent of deterioration are usually extremely variable, often depending upon unforecastable factors such as conditions of use of each item, and the effects on reliability and durability are also highly variable. For example, the scatter in times to failure due to material fatigue for carefully controlled, nominally identical items subjected to identical stress cycling typically varies by a factor of 10. Fatigue life predictions for new designs must take account of further uncertainties in stress cycle application and product variation. Yet fatigue is probably the best understood, most predictable type of deterioration encountered in engineering.

As a result, these aspects are often difficult for designers to assess fully or accurately, and CAE software provides very limited or no help*. Therefore the only practicable way to determine the likely effects of factors such as these is to plan a test programme, in which a sensible sample of the product is tested and the results observed.

Variation

The same arguments apply to analysing and testing to determine the effects of variation in parameters such as dimensions, strength, electrical properties and any others that might affect performance,

---

* Software for fatigue life assessment is available, using FE models and materials data. See the homepage of [12].

reliability and durability. Such variations can exist initially as a result of tolerances, manufacturing variations, etc. They can also develop over time, due to, for example, electrical parameter drift or mechanical wear. Analysing and testing for variation will be covered in Chapter 10.

OPTIMISING THE DEVELOPMENT TEST PROGRAMME

Product testing can be usefully considered to be in two categories. In the first, we hope that the product will not fail. A test to demonstrate that a design will comply with the specification, for example that it will not consume more than the power specified or that it will comply with a mandatory safety standard, should be planned on the expectation of success. On the other hand, tests to provide assurance of strength, reliability or durability must be planned to generate failures, or at least evidence that failure will not occur, or will be very unlikely to occur, within the product's expected conditions of use and lifetime. The uncertainty and variability inherent in forecasts of such properties mean that testing to generate failures outside the operating regime is usually the only practicable way of providing assurance that failures will not occur inside*. Therefore the test programme must distinguish between the tests that should demonstrate success and those that are planned to generate failures.

It is not necessarily the case that the whole product must be tested to failure. At one extreme, a civil engineering structure such as a bridge cannot economically be tested to failure in fatigue due to cyclic loading. Therefore the bridge designer uses the knowledge of tests to failure of samples of the bridge steel, and includes strength margins that guarantee that such failures cannot occur. On the other hand, the designer of an aircraft wing or pressure cabin is denied the luxury of such large safety margins, and so must create a compromise between weight and safe life (as well as other features such as ease of crack detection, provision of crack stoppers and minimization of stress concentrations). Such a compromise must be tested to failure to provide assurance that the inevitably uncertain analyses are sufficiently conservative.

For complex products, which can fail in many different ways, the test plan must create the likelihood that as many as possible of these will occur, so that action can be taken to prevent or minimize their

---

* The same logic applies to testing to assure safety. The main difference is that higher levels of assurance are required if failure can cause hazards.

occurrence. A new design of photocopier is a good example. Such a product contains a large variety of electronic circuitry (power, analogue, digital, high voltage), sensors, mechanisms, drives, controls, displays, etc. To be competitive it must have attractive performance features, and it must be reliable. It must also comply with standards for safety and electromagnetic compatibility. It is inconceivable that a design team, however skilled and experienced, could create a design for a new copier that would meet all of the performance, safety and reliability requirements without testing to find the shortcomings. For such a product the test programme will have to be long, intensive and expensive. There will be surprises and disappointments, and the tests will not run to the initial plan or budget. The reason for this is the uncertainty involved: we cannot know what failures will occur, or how often, or how much effort and time will be needed to correct them. Because the test programme is expensive there will be pressure to reduce it, or at least not to extend it. On the other hand, the engineers involved will be concerned that, by the end of the planned test programme, there will still be features that could benefit from further testing and development, and failures will have occurred which cannot be guaranteed not to recur in service.

There is no way of avoiding this dilemma for complex products such as copiers, cars, trains, and planes. The effectiveness of the test programme in providing assurance and generating improvement, and the cost impact, can be optimized by concentrating on two principles of product design and development:

- Ensuring that the engineers concerned have the knowledge, experience, facilities and leadership to create the maximum proportion of designs that are right.
- Planning a test programme that maximizes the chances of revealing all designs that are not.

Costs and Benefits

The major contributors to the cost of testing are the articles being tested and the manpower cost (people and time). Test facility costs can also be high, and for these time also is a multiplier. Therefore we must obtain the maximum information using the minimum number of items, for the shortest practicable time.

However, it is important that the test programme is not considered to be a burden on the project. Rather, testing is an investment in the future, to minimize downstream costs of failures (re-design, factory scrap and re-work, warranty, reputation, etc.) These costs can be very high, and often greatly exceed the amount invested in testing (remember the "times ten rule!). Therefore the development test programme should be considered as a value-adding activity, to be optimized over the life cycle of the project, not merely as costs to be minimized in the short term.

Quantity to Test

The number of items to be tested will obviously depend upon their costs. Every prototype aircraft or spacecraft represents a huge additional cost to the development project, so the number to be provided for test must be minimized and the test programme must be carefully planned to make the best use of these.

For products such as engines, cars, copiers and electronic test equipment, the cost of prototypes is high, but not so high that the number of prototypes to be used in the test programme should be cut to a bare minimum, since the cost of insufficient testing or late completion of testing is likely to be much greater than the subsequent costs of undetected and uncorrected problems. Obviously, for low-cost items (fasteners, electronic components, etc.) the test programme should not be constrained by the cost of hardware provision.

When to Test

If testing is delayed until the designs are considered to be complete and nearly ready to be released for production, then problems discovered during testing at such a late stage could seriously disrupt the overall project programme. Therefore the correct approach in any development programme is to start testing as soon as hardware can be provided that is even only partly representative of the expected final design, accepting that the designs might not be fully representative at this stage, in order to discover problems as early as possible. This early testing must then be followed by repeat tests of the improved designs to demonstrate the effectiveness of the improvements and to show up any further problems.

Testing Systems and Subsystems

When a product consists of several sub-systems, it is necessary to consider testing at different levels of assembly. Testing of lower level components and sub-assemblies is less expensive and can be more effective in exploring the limitations of the designs than testing at the level of the complete system. For example, the performance and reliability of an electronic engine fuel control unit can be explored using several units, operated over a wide range of conditions, far more cheaply and effectively than by reliance on testing of the vehicles in which the units will operate. Lower level tests can also be carried out earlier than system-level tests. Of course, most systems comprise a range of components and sub-assemblies, ranging from simple to complex, mature to innovative, and standard to unique to the new system being developed. Items that are unique, innovative, or complex must be given priority, since they obviously represent the greatest risks. When items are bought from external suppliers the system developers must ensure that the tests applied are adequate, either by reviewing the suppliers' test programme and results, by conducting separate tests, or a combination of both.

Component and sub-assembly testing provides assurance that the products concerned can operate under the test conditions applied. However, the test conditions can differ in important and subtle ways from the actual conditions in the system. For example, the way that a sub-assembly is fitted into a system might induce static and dynamic mechanical forces that are different from those applied in tests of the sub-assembly. Equally important, system-level interactions might occur that are not simulated in lower level tests. Typical of these are the effects of location and connections on electromagnetic compatibility and timing of electronic units, and the interaction of engine and transmission dynamics on vehicle noise and vibration.

Finally, it is not normally possible to confirm the performance of a system purely by synthesizing the performance of its components and sub-assemblies, due to the many uncertainties and interactions that exist. Therefore it is essential that the complete system is tested, as well as its components and sub-assemblies.

Simulation

As discussed in Chapter 7, it is possible to use CAE simulations of designs to test their performance. Using capabilities such as these to test the performance of designs can obviously be much less expensive than making and testing hardware. The testing can also be started earlier.

The use of computerized simulation methods for testing engineering designs presents opportunities for the creation of better, more economic products. However, CAE cannot replace or reduce the need for engineers to have knowledge of the principles involved and experience of their application. Rather, they should have enhanced awareness, since simulations might not reveal the truth, whereas real tests always do.

No simulations can determine the reliability or durability of a design, in the sense of predicting how or when failures might occur, except in some very limited cases. Other limitations were discussed in Chapter 7. Therefore, while simulations can be very helpful in creating correct designs, they cannot replace the need for effective real testing.

TEST CONDITIONS

Most engineering products must operate in a range of environmental conditions. These can include temperature extremes and changes, vibration, shock, humidity, corrosive atmospheres, electrical power conditions, etc. In particular cases other conditions might be important, such as storage and transport, altitude, solar or other radiation, acoustic noise, electromagnetic interference, etc. The environmental conditions can influence performance, durability and reliability, so their effects must be considered in planning the test programme. The extreme values of the environmental conditions, the times and cycles over which they occur and their rates of change must all be considered. Environmental effects are often interactive, so that the effect of combined environments and operating stresses can be more severe than any one effect in isolation. For example, the growth rate of fatigue cracks in materials such as steel can be accelerated by the presence of corrosion, and the quality of electronic soldering processes can be affected by combined factors such as solder temperature, composition, preheat and several others. Therefore the test programme must include the real, combined environmental conditions as far as is practicable. The tests should apply these conditions at levels higher than expected in service, as described below.

The human aspects of the environment are also important, since people, working as producers, packers, shippers, installers, operators and maintainers, all influence the quality and life of the product. Therefore the test programme must also include the human aspects, not only operating and environmental stresses.

ACCELERATED TEST

Since both test time and units to test are major cost considerations in almost any development programme, it is essential that the right balance is struck between investment in testing and the payback in terms of increased reliability and durability. The most effective way to economize on test time is to apply accelerated testing. This involves the application of operating and environmental stress conditions, including cycling frequencies where appropriate, higher than those expected to be encountered in service. One form of accelerated testing is step-stress testing, in which the stress levels are progressively increased until failure occurs. This is also called highly accelerated life testing (HALT). Accelerated tests can greatly increase the cost effectiveness of a test programme. However, it is necessary to investigate failures carefully to ensure that they are relevant. The criteria of relevance must be the possibility of occurrence in service and the evidence presented of design safety margins. In situations where the relationship between stress intensity and failure is fairly well known, for example in fatigue, these can be used to evaluate the reliability at lower stress levels. However, such relationships are uncertain and should be used cautiously.

The logic that justifies the use of very high "unrepresentative" stresses is based upon four aspects of engineering reality:

1. The causes of failures that will occur in the future are often very uncertain.

2. The probabilities of and durations to failures are also highly uncertain.

3. Time spent on testing is expensive, so the more quickly we can reduce the uncertainty gap the better.

4. Finding causes of failure during development and preventing recurrence is far less expensive than finding new failure causes in use.

It cannot be emphasised too strongly: testing at "representative" stresses, in the hope that failures will not occur, is very expensive in time and money and is mostly a waste of resources. Despite this, accelerated tests are often applied inadequately or not at all.

Design engineers are often reluctant to agree to testing of early prototypes at conditions that exceed those specified, and they might consider failures of such early models generated under such conditions to be unrepresentative and therefore not worth investigating or taking action to prevent. However, to repeat: the criterion for relevance and action must be whether the failure could occur on production items during their service lives.

There is a further benefit in the use of the accelerated test approach: since the test duration is short and failures will be generated in that time, the designers can watch how their creations respond to the stresses applied. This can be a very powerful and effective learning experience, leading to rapid design improvement and, sometimes even more valuable, to improvements in the designers' capabilities to create better future designs.

TESTING SOFTWARE AND DIGITAL SYSTEMS

A newly written software program is likely to contain errors. Errors are created by the people who produce the specification and by those who interpret it and write the program. If a program is simple, the kind that one person conceives and writes, then there is less scope for the creation of errors than for a large program involving complex specifications, several programmers and large amounts of code. For example, it is inconceivable that the software for an aircraft flight control system could be written without any errors. In software-controlled systems such as this, and others such as railway signalling or machine control, the consequences of errors that remain in the software can be serious. Even for simpler, less critical systems errors can be expensive and embarrassing. Therefore it is important that any errors that are created are detected and corrected. In Chapter 7 we discussed how this should be done as part of the design process, but, as with hardware, design analysis cannot be assumed to provide complete protection against errors, so the software must be tested.

Software cannot degrade or change in any way, so if a program works once in a particular combination of circumstances it will always do so. Conversely, if it fails once in particular circumstances it will always do so. Therefore software testing is not concerned with testing over time, but with testing to cover all possible situations in which errors in the program could cause the system to fail. For example, a process control program in which a measured value such as an electrical resistance is used as a divisor would stop running if the value measured was equal to zero, within the precision of the measurement, since dividing by zero gives an indeterminate result. However, the program would continue to run if appropriate code was included to protect against zero being used as an input to the divide operation. There are of course many other types of error that can be made in software writing, and they can be difficult to detect by inspection or analysis. The testing must therefore be planned to cover the possible situations in which errors could cause the system to fail.

In theory, the number of different situations is 2^n, where n is the number of separate logic states (total of loops and paths) that the software can experience while operating. Of course, for a typical large program n can be many hundreds or thousands, so the time to perform such a number of tests is not realistically available. It is therefore necessary to plan a selection of tests that cover the important or critical situations, as well as the more normal operating conditions. There can be no "scientific" or mathematical way of doing this, since the likelihood of situations and the effects of failures can be judged only by the people involved in the design of the system. It can be extremely misleading to assume that all situations are equally likely and that all failure effects are equally serious. On the other hand, errors in software are invisible to the operator, and failure occurs without any warning.

As with hardware testing, all tests performed must be carefully recorded, including the test conditions and results. If errors are detected they must be recorded, corrected, and the system re-tested. The test record is the only evidence that the software will work correctly under the test conditions specified.

Despite these obvious truths, attempts have been made to "measure" the reliability of software by recording numbers of failures and performing analyses to determine the likely number of remaining errors, or the likely rate at which they will cause failures in the future. Expressing software reliability in these ways is not meaningful, particularly in engineering applications. A program that fails ten times but with unimportant consequences is not necessarily less "reliable" than one that fails only once with serious consequences, or that contains only one error that can cause a hazardous failure under certain conditions.

Testing the designs of digital logic systems presents the same basic problems as testing software. However, modern electronic design automation software does provide the capability of testing the design over the full range of logic states, since digital hardware contains only a small number of basic logical functions. The correctness of the designs of standard components such as microprocessors and memories can normally be taken for granted, but the designs of digital systems built around them, and of application-specific integrated circuits, must be tested.

Since digital hardware is manufactured using processes that are variable, rather than merely copied as is software, all manufactured items are not identical. Some might contain defects that prevent operation, under all conditions or under particular conditions. Therefore each system produced must be tested for correct function. This testing need not explore the design boundaries of the system, since production variation that can affect performance can almost always be detected by less exhaustive tests. For example, simulation and testing of a new standard digital integrated circuit takes hundreds of hours, but each production device would typically be tested for not more than a few seconds.

TESTING THE PROCESSES

For every product that is to be manufactured in quantity, it is imperative that the test programme includes testing of the manufacturing processes, not only of the product design. We discussed the importance of design for manufacture earlier: testing for manufacture is equally important. The more that the costs of the manufacturing processes influence the product costs the more important it is that the development test programme includes testing of processes.

Manufacturing process costs depend upon the difficulty or novelty of the processes, the number of processes and the cost of the production facilities. The cost of facilities (jigs, automation, test equipment, etc.) will depend upon the product and on the quantity to be produced. Normally, the larger the quantity to be produced the more need there is to invest effort and resources in minimizing the marginal cost of production by the use of capital facilities, since the fixed costs can be amortized over large production quantities. However, when such facilities and the methods by which they will be used are determined as part of the overall design and development programme, then their effectiveness must be tested. Unfortunately it is not uncommon for products to be designed and developed to the point at which drawings and process instructions are released for manufacture, and then the production people are expected to begin manufacture to schedules and budgets determined during design. Of course such schedules and budgets assume that there will be no problems, even if they do allow for initial learning by the production people. However, process problems always occur. The only way of avoiding them during the manufacturing phase is to test the production processes during development so that they can be improved and refined before production commences. Problems can be removed by changing the process or by changing the design, or possibly both together.

As with product testing, process testing must begin as early as possible to give time for improvements to be developed. As far as practicable the processes should be performed by the people who will operate them, using the equipment (tools, jigs, test equipment, etc.) that will be used in production. If the processes are performed by development engineers, using unrepresentative equipment, it is very likely that real process problems will not be detected. The production people must be integrated into the development test programme at a sufficiently early stage, and their findings and recommendations must be given as much attention as those related to product design aspects.

Maintenance, including servicing and repair, of the product in service also involves processes, many of which are the same as or similar to the manufacturing processes. Therefore the product support people must also be involved in the development test programme and maintenance operations must be tested and refined.

Integrating production and maintenance process testing into the development test programme can generate considerable benefit when the product enters the production and use phases. The initial cost per unit produced will be reduced and the subsequent cost reduction rate due to learning effects will be increased, since fewer problems will be encountered and there will be a core of trained, experienced production people at the start. The same applies to the maintenance aspect, as the service people will also have benefited from the experience gained during development testing. The experienced production and service people will be better equipped to train others. Also, the need for costly changes to designs and processes, particularly changes that delay production, will be greatly reduced.

COLLECTING AND ANALYSING TEST DATA

Obviously a development test programme must be carefully organized and run in a disciplined way to ensure that all necessary tests are performed, that tests are not unnecessarily duplicated, and that the maximum information is obtained within the constraints of time and cost. An essential aspect of this discipline is the full and accurate recording of all test results. It is important to investigate every failure that occurs, under any conditions. Therefore disciplined, 100% failure reporting and investigation must be imposed from the start of the test programme. For every test it is necessary to record details of the test conditions, the design status of the items tested and results of successes and failures.

Recording success is usually relatively easy. Recording failures is often considered more of a problem, so they might be left unreported. However, every failure shows a potential or actual weakness and consequently an opportunity to improve the design. It is therefore essential that the test reports include details of all failures, preferably on separate failure reports. Action should then be taken on these reports to investigate fully the causes of failure, to identify the corrective action necessary to prevent repetition and the testing necessary to prove that this is effective. The data on failures and corrective action should be reviewed by the project leader, and he should ensure that the improvements shown to be necessary are implemented and tested.

Failure data collected from development testing should never be used to demonstrate, in a directly quantitative way, that reliability is being improved as a consequence of fewer failures being observed. Since the objective of much of the test programme is to generate failures in order to identify opportunities for improvement, the mere counting of failures and using them as an index of reliability is misleading. It is also counter-productive, since it discourages the reporting of failures. Also, the conditions under which products are tested in development are usually different from the conditions experienced in use, particularly if accelerated tests are applied. Therefore the only correct indicator of reliability improvement during development is the rate at which failure modes are discovered and corrected.

CONCLUSIONS

Testing is usually the most expensive and difficult task in the development of engineering products and systems. It is in the test phase that the errors and oversights that are inevitable in nearly all modern engineering design should be detected and corrected, rather than coming home to roost after the start of production and introduction to service. Development test programmes always entail a large amount of uncertainty, because we cannot predict what problems might occur or how serious they might be. We can greatly reduce the potential for errors and oversights by ensuring that design teams are better trained and more experienced. We can also detect and correct design problems by effective use of design analysis and review methods. However, it will nearly always be necessary to perform tests to ensure that the design of the product and of the processes is acceptable in terms of performance, reliability, durability, safety and, in many cases, regulatory compliance.

The technology of test is complex and dynamic. Few engineers are adequately trained or experienced in test methods and economics, and inappropriate practice is very common (some examples are described in [14]). In many cases improvements in test management and methods, as described in [14], can generate large savings and commercial advantage in terms of engineering and manufacturing productivity and product reputation.

Managing the test programme is a difficult and demanding task, the management and performance of which can greatly influence the outcome of the project. Testing must therefore be organized and managed so that the design and development teams work as one, with product and process designers closely involved in planning and conducting the tests, and the other engineers who contribute to planning and conducting the tests must be encouraged to contribute to the design process.

9

PRODUCTION

PRINCIPLES OF PRODUCTION

The first principle of engineering production is to convert designs into products, at the lowest cost. The design represents ideas and provides data and instructions for production. The cost of production is largely determined by the design.

The second principle of engineering production is that all processes are operated or influenced by people. Automation has replaced human effort in many tasks, particularly in mass production, but no complete engineering manufacturing process has been or is likely to be fully automated in the sense that designs and raw materials are input at one end and products come out the other. Processes that have been automated still depend on people for planning, maintenance, inspection of finished products, repair, and often for feeding work in and out. The design of the product is, like a music score, the unique statement of the designer's intent. Each product made to the design is also unique, like a performance, but is an interpretation of the design. To varying degrees designs, and the systems of production that are used, allow freedom of interpretation in manufacture. Ideally, for items made in quantity, we want all interpretations to be identical, with no variation in the items produced or in the methods used. However, this is never possible. As will be explained in Chapter 10, processes inevitably generate variation, and this is increased by human involvement. Therefore items can be produced which are imperfect or defective. The extent of variation and the proportion that is defective affect productivity and costs, and in the case of most engineering products these costs can be very high.

The third principle of production is that, as far as practicable, nothing should be made that cannot be billed immediately it leaves the factory. Ideally, every item made should have a customer. This is not difficult for large capital projects and other very expensive items such as aircraft, military systems and spacecraft. Such products are usually ordered before manufacture, so meeting this principle presents no problem. However, most engineering products are manufactured for an expected market, and this must be forecast. The main reason for manufacturing to meet forecast demand is that manufacturing and shipping take time. If it were possible to manufacture and deliver very quickly on receipt of orders there would be less need to create unsold stock. Speed of response also relates to the time it takes to commence manufacture of new or improved designs or to satisfy particular customer requirements.

Productivity therefore depends upon how much is made per unit of input (money, effort, time), how much is made correctly, that is, with minimal variation or error, and how long it takes to sell the products.

Manufacturing productivity is a key contributor to competitiveness, since it greatly influences the price that can be charged and the profit that can be made, as well as the customers' preference for the product in terms of delivery and quality. Manufacturing productivity is therefore a major factor in determining market share and growth.

SYSTEMS OF MANUFACTURING

The first engineering manufacturers, such as the makers of early clocks, watches, engines and motor vehicles, up to as recently as the early twentieth century, used the craft approach to production. Individual craftsmen, or possibly a small team working for a master, would perform all the processes of cutting, shaping, assembling, adjusting and testing, working on individual products from start to finish. Each product was unique in the sense that parts were not interchangeable. The craftsmen understood all of the processes and how the product worked. Often, for example with clocks, the craftsman was also the designer. Great skill was necessary, and nearly all products were made to order. Therefore engineering products such as these were very expensive and were built in small quantities.

Eli Whitney, the American inventor of the cotton gin, was the first to demonstrate the principle of mass production in the manufacture of muskets for the American army in 1798, by having men make interchangeable parts using templates and machines, rather than each rifle being an assembly of uniquely crafted parts. The essential feature of mass production was the replacement of craftsmanship by a large number of simple tasks that could be performed repetitively by unskilled workers with minimal training. For the system to work it was necessary for all components to be interchangeable, so that the assembly workers did not have to make any decisions or adjustments, and all tasks could be completed within the short times allowed. The monotony and discipline of plantation work was being reflected in the new industries: Whitney had grown up in pre-Civil War cotton plantation society, and slavery was not far from peoples' memories.

Henry Ford refined the idea of mass production in 1908, with the objective of producing the Model T Ford car in large quantities and at low prices. Soon afterwards the refinement of a moving track was added so that the work came to the workers, thus saving the time involved in their moving from place to place.

Mass production led to enormous gains in productivity in nearly all engineering industries, and it enabled mass consumer markets to be developed in motor vehicles, electronics and many other products. It is the system generally in use today, with variations between different industries.

PROBLEMS WITH TRADITIONAL MASS PRODUCTION

Despite the great success of the mass production approach, it contained features that limited its potential, particularly in modern engineering. The main problem was that it created a fundamental separation between production workers and the managers and specialists such as engineers, scientists and marketing people. Two cultures were created: on the one hand the "workers" or "operatives", who were told what to do and how to do it, and, on the other, the "professionals", who planned, organized, solved problems and made decisions. There was nearly always a sharp difference in educational attainment, and this accentuated the division. The result was that the two sides were motivated differently. The "professionals" generally were motivated towards the objectives of the business. The "workers" were motivated towards protecting their interests, which in important ways were often in conflict with those of the business. The "workers" wanted security and to maximise their incomes. On the other hand, they were largely denied the opportunity to satisfy the higher needs of esteem and self-actualisation in relation to their contributions to the business. The history of alienation between "management" and "workers", and the growth of trade union influence in manufacturing industry were direct and inevitable consequences. Enlightened managers and trade unions did much to mitigate the effects of these disadvantages, but the system was inherently unstable. Any perceived recalcitrance by either side could lead to suspicion and loss of cohesion in what should be the business team, or even to conflict.

Mass production is inherently a complicated process. Many parts, materials, and sub-assemblies must be brought together, at the right time and to the right place. These must be balanced against the production demand and the capabilities of the various processes. For example, in car production, sheet steel blanks must be fed to body panel presses, then the panels are moved to the right parts of the assembly line, and fittings such as transmission units, seats, etc., must all be delivered. When production involves machining and measuring, the various processes must be balanced against machine availability and throughput. These aspects of production engineering led to the development of planning techniques such as line of line of balance (LOB) charts and computerized planning systems such as Manufacturing Resources Planning (MRP). Production planning departments managed the specialized activity, and suppliers were managed by other departments with names like Purchasing or Materiel.

In order to ensure that materials, components, and sub-assemblies were available when needed, suppliers were required to deliver in advance, and stocks were held near to the production line. The amount of stock held depended upon the production rate, the delivery time and the perceived risk of non-delivery. Obviously this stock could represent a very high cost, since it must be paid for some time before its cost could be recovered by sale of the end product. Determining the optimum stock levels was therefore a critical function, but it could also be very difficult and uncertain.

Suppliers to mass production lines must be encouraged to keep their prices as low as possible. Typically they would be given specifications, or orders would be placed against their own specifications, as determined by the product designers. Selection was based upon the lowest cost that could be negotiated by the purchasing department, and whenever possible several suppliers would be used and played off against one another in order to maintain pressure on prices, delivery and quality. Suppliers in turn would adopt strategies that were optimal for them, but which might not be optimal for the producer. For example, they might deliver items that fell below quality standards, hoping that the problems will not be detected. Suppliers were often not involved in the design process for the final product nor in production planning, so valuable contributions they could make were omitted. Finally, since suppliers knew that the purchaser might at any time change his sources, they would attempt to build this risk into their prices, and they would be unwilling to share engineering or cost information that the purchaser might use against them, for example by passing to a competitor. The supplier-purchaser relationship was one of the most difficult problem areas in the management of mass production. It was often adversarial rather than cooperative.

The more that a mass production process is capitalized and automated, the more difficult it can become to change the product designs or to increase the variety of the products. However, modern markets for most engineering products demand rapid changes to accommodate new designs and variety, such as different features and styles. Process changes and variety further complicate the planning process, and add to the difficulty of demand forecasting.

The system placed great demands on managers, who were expected to plan and coordinate every task whilst maintaining worker motivation and trust. Supervisors were usually appointed to provide on-the-spot management, and their role was often ambiguous: were they junior managers or promoted workers? Production managers were, in turn, often created by promoting supervisors.

The difficult task of managing the mass production system was not effectively taught as part of most engineering graduate level courses, which tend to concentrate the more "scientific" fields of theory and design. Graduate courses in production engineering are not considered as having the status of the more conventional engineering courses. It is therefore not surprising that production engineering has come to be seen as of lower status and esteem than design, and few engineers make the transition either way.

Mass production was clearly strongly influenced by F.W. Taylor's principles of scientific management. The design of the product and of the manufacturing processes, to enable the whole manufacturing task to be broken down into a series of quick, simple tasks, then planning and implementing the system and finally making it all work were tasks for managers and specialists. The people who actually performed the manufacturing tasks did not need to understand the design or the processes; they needed to know only how to perform their small individual contributions.

Since the tasks were made simple and repetitive, a natural later development was automation. Most modern mass production lines, particularly car assembly and electronics manufacture, utilise very high degrees of automation for tasks such as machining, welding, painting, component insertion, etc. However, human inputs are always part of the process, and the efficiency and reliability of the automation depends to a large extent on human performance. The nightmare vision of mass production and automation completely dehumanizing the workplace, even the people in it, as shown in the Charlie Chaplin film *Modern Times*, has not generally materialized, since managers have realized that a balance that must be made between total automation and job de-skilling at one extreme, and craftsmanship at the other. Nowadays most mass production operations take account of the contributions that workers can make to improvement and efficiency, production workers are often involved in planning, and the work is made more interesting by allowing the workers greater responsibility, for example by job rotation and by performing inspection and other functions. The lessons of Hawthorne and the teachings of Maslow, Drucker, Deming and others, discussed in Chapter 3, have been applied to develop and improve the human contribution to mass production.

Mass production used to be considered synonymous with mediocrity, as craftsmanship was with quality. Quality of production is always relative to expectations, and it is possible to ensure that the output of a mass production system attains high quality. However, the system made this more difficult and expensive. As described above, it is difficult to motivate the workers to produce only perfect work, and if they make mistakes these can be detected only by subsequent inspection, and therefore at additional cost of inspection, work in progress and repair or scrap. Fixing mistakes downstream involves further costs, which can be high if disassembly, reassembly and re-test are needed. Inevitably a proportion of mistakes will not be detected, and these could result in warranty costs and lowered product reputation. The importance of quality as a factor in competitive markets has been one of the major factors that have led to reappraisal of the mass production concept.

Modern competitive markets also demand that new products and variety within product ranges, often involving new technologies, components, materials or processes, can be produced and marketed quickly. High standards of quality must be achieved from the start of production or the advantage of early introduction will be degraded and the product's reputation will be damaged. Beating the competition to market with innovative, high-quality products provides great strategic advantage. However, attempts to do so obviously carry considerable risk, and the production system used must minimize these. To meet these severe demands, and to counter the disadvantages of traditional mass production, a new philosophy of production has been developed.

THE NEW PRODUCTION PHILOSOPHY

The new philosophy of mass production was described by Drucker in *The Practice of Management* [1]. Drucker called this "new style" mass production, the main difference from the traditional approach being the capability of the system to create a wide diversity of similar products by using different but standardized parts from which different products can be assembled. He emphasized that this approach would place new demands on managers and on workers. In particular, there would be an absolute necessity for an integrated approach to design, development, production and marketing. Also, the assembly workers would have to be skilled and versatile.

The person most closely associated with the early practical application of the new approach is Taiichi Ohno, Production Manager of the Toyota car company in the 1950s. He observed the mass production methods then being applied by American car manufacturers*, perceived the problems, and (with Shigeo Shingo) implemented the new philosophy. The "Toyota Production System" (TPS) enabled them to outperform their competitors in world markets. It was quickly copied by other Japanese car manufacturers and by companies in other industries***. The concept has since spread to most of the industrialized world, but it is not easy to apply it, for reasons that will be discussed later.

The development of the Toyota Production System and its application in the world automotive industry is described vividly by James Womack, Daniel Jones, and Daniel Roos in *The Machine That Changed the World* [8]. The authors gave the name "lean production" to the new philosophy. Much of what follows in this chapter is based upon their book. The new approach to manufacturing is as much a revolution as was the introduction of mass production. It shares some of the features of mass production, particularly in the use of

---

* Ohno and other Toyota executives visited General Motors as part of American help to the post-war Japanese economy. They were permitted to observe GM methods and noted what they saw. Years later, after Japanese companies had overtaken their Western competitors, Ohno explained that on their return to Japan they did not apply the GM methods in an attempt to catch up; instead, they decided that they had learned how NOT to build cars if they were to overtake the West. Note also: Toyota did not send a team of junior "experts". The responsible top managers came, saw and conquered.
** See the quotation at page xiii.

automation, and the casual visitor to a factory employing the new approach would not detect the differences. However, it largely removes the problems associated with traditional mass production and it has been the main reason for the dramatic improvements in cost and quality of products such as cars, white goods, electronic components and equipment, and many other products.

There are several features that distinguish the new philosophy from traditional mass production. The fundamental difference is in the way that the contributors to the process, the production workers and the suppliers, are empowered and treated. Instead of being considered to be providers of labour and components, with little or no contribution to make to the processes of design, planning and improvement, they are made active members of the total production management team. Of course this philosophy is consistent with the modern theories of motivation described in Chapter 3.

Workers are organized into teams and are trained in multiple skills. They are all treated as skilled people whose capabilities can be continually developed, and who have the intelligence and motivation to contribute to the improvement of the processes and the business. The team members can decide among themselves what jobs to do, usually under the overall control of a supervisor. The supervisor is responsible for ensuring that all jobs are performed and for training and helping the team. For example, if a team member is absent the supervisor will know which other members are trained to perform the tasks, so that the production process is not interrupted. Typically, each working day or shift begins with team meetings at which the supervisor briefs the team on the plan for the day, any problems are discussed and work assignments are agreed.

Continuous improvement of all processes is a fundamental principle of the new philosophy. Improvement efforts are directed at eliminating "waste" (muda), which includes inefficiency, defects and failures. The Japanese word for this is Kaizen. The principle is different to the approach to improvement common to most traditional management thinking. Instead of improvements being generated by large, planned, management-imposed changes, with justifications presented with formal cost-benefit analyses, everyone in the process is expected to contribute to continuous improvements in quality and productivity. Of course major improvements are not excluded, but the steady accumulation of small improvements generates greater overall improvement than the occasional attempt at large steps imposed from above, which can be destabilizing and counter-productive. Since the improvements are generated by the people most closely involved, they are more likely to be effective and to be incorporated into the system. Also, there is far less danger that the results of changes will be counter-productive, and if they are it is easy to revert to the previous method. Therefore problems, mistakes, and defects are reported immediately, because they all represent opportunities for improvement. No stigma is attached to reporting them. The workers and the teams are given considerable management authority over the running of their part of the process: they can stop it if they have problems, but they are also responsible for its performance and improvement.

The workers and the teams are totally responsible for quality, as well as for the quantity of work produced. Quality is given priority, so that no defective work is permitted to be passed on to the next production stage, even if this means stopping the line. Every worker has the authority to do this, and no provision is made for re-work at the end of the line. There are no separate inspectors: each worker inspects his or her own work, and also the work of previous workers in the line. Their training and skills enable them to carry out inspection and minor repairs or adjustments if necessary.

The transfer of such power to the process workers demands considerable trust from higher management. This trust can be justified on the basis of the motivational principles discussed in Chapter 3. However, it also requires that management is diligent in selecting and training the production staff and in taking action on ideas they generate but which require higher management action, for example a change to another part of the process or to a supplier's component or process.

The payment arrangements for production workers must support the philosophy. Payment must be related to overall performance and not linked merely to output. Pay is usually linked to experience and training, so that there is clear recognition that workers' capabilities and potential to contribute depend upon increased skill and knowledge.

The book by Masaaki Imai, *Kaizen* [16] describes the approach. The methods taught and used for improvements in quality and reliability are described in the following chapter.

Suppliers

The suppliers to the new production system are also treated as partners in the enterprise, in terms of trust and responsibility. The selection of suppliers must start early in the design phase, as described in Chapter 6, because they will be given responsibility for ensuring that their products meet all requirements of suitability, quality, price and delivery. Therefore suppliers must be selected primarily on the basis of reputation, and not on the basis of costed bids against detailed specifications as is the usual practice in traditional production systems. The partnership extends from design into the production phase and is a long-term one, not subject to change or cancellation by the purchaser unless there is serious failure to perform. Problems that do occur are worked on together; the purchaser will help or train the supplier to solve problems and improve his processes, as he would do for his own production workers. In the same way that an employee should be considered as a long-term asset, to be developed and employed on future work, so suppliers are treated as long-term partners, not merely as suppliers to a particular product. They are given training when appropriate, and encouraged or required to use compatible methods and systems such as computer-aided engineering.

Two features of supplier contracts in the new production system have been widely publicized: the single supplier agreement and "just-in-time" production.

Since suppliers are selected on the basis of reputation and trust, it follows that it would be inappropriate to use multiple suppliers for the same product or service and to force them into price reductions or other improvements by threatening to switch orders. This is, of course, the traditional mass production approach. Instead, suppliers for particular products or services are given sole contracts: they compete for the contracts, but not for subsequent business. In fact, contracts are sometimes placed with two or more suppliers, and then the business is shared between them, with proportions being adjusted in line with performance.

Suppliers are required to improve their processes and service continually and to reduce their prices progressively. This is consistent with Deming's teaching and the *kaizen* approach, and is achieved by close cooperation between the purchaser and the suppliers. Prices are not reduced in order to reduce the suppliers' profits, but to maintain competitiveness of the end product; this is seen to serve the suppliers' interests, particularly in the long term.

For such close, long-term relationships to exist it is necessary for both sides to share information on plans, methods and costs. The supplier needs to know the purchaser's production plans and methods in much more detail than is expected in traditional mass production contracts. In turn, the purchaser needs full data on the supplier's methods and costs, so that cost reductions can be developed in cooperation. The supplier can hand over this type of commercially sensitive information since he has confidence in the long term, multi-project nature of the contract, and he knows that the information will not be passed on to competitors. The purchaser-supplier relationship is cooperative and trusting rather than adversarial and secretive, and suppliers are given much greater discretion within the general requirements of their contracts. The principles of human motivation, described in Chapter 3, are thus extended to the suppliers, from their senior managers downwards.

"Just In Time" Production

"Just in time" (JIT) production replaces the concept of sub-assemblies and components being delivered in advance and then being fed into the production line, by a system that requires suppliers to deliver each item, or small batches of items, directly to the point of assembly exactly when needed. The system works by replacing the conventional mass production approach of forward ordering of components to provide buffer stocks, with ordering of components only when needed. The production processes "pull" supply, rather than supply being "pushed" by the production forecast. The system also operates internally, from process stage to process stage. Manufactured items are moved forward only when the next production stage indicates that it is ready to use the item, and the preceding stage will stop rather than produce work that is not wanted. The notification that items from previous stages or from external suppliers are needed is by use of the *Kanban*, which is a simple form, often accompanied by the appropriate packaging for the items required.

In the JIT system the manufacturer is relieved of the cost of stockholding, and there is a great reduction in the cost of work in progress, since the amount of partly completed product is minimized. JIT goes much further than this, however. The purchaser performs no separate inspection: the production workers inspect the items as part of the assembly process. If defective items are delivered, there is no backlog of defective stock. Instead, there is immediate feedback so that corrective action can be taken before more defectives are delivered.

JIT can work effectively only if very high quality standards are maintained, so that defective deliveries are rare. The purchaser must work closely with the suppliers to identify sources of problems and to correct them. There is no slack in the system to allow correspondence or negotiation. Action must be immediate, flexible and face-to-face, with both sides committed to solving problems as quickly as possible, since both production lines will be stopped until solutions are developed and put in place. The action takes place between the people most directly involved, i.e. production and design engineers, not between purchasing and contracts department staffs.

JIT supply is obviously easier when the suppliers' production lines are close to the purchasers. However, it is left to the suppliers to make arrangements for transport and intermediate storage, though the purchaser should assist and advise as part of the general cooperation. Ideally the supplier will also manufacture just in time, as each order is received, and will place orders on lower tier suppliers in the same way, so that the concept is deployed downwards through the whole manufacturing and supply chain, as far as is economic and practicable.

Since JIT replaces much of the need for relatively long-term forecasting of every detail of the production operation with a more general forecast, and then the "*Kanban*" system determines detailed requirements, the need for separate planning staff and complex computerised planning systems such as MRP II are removed. All of the people in the operation are involved in planning, operation and improvement.

MAKING THE TRANSITION

Implementing the new production philosophy is easy in principle but difficult in practice, particularly within an existing organization or with existing products. Many attempts have been made to graft elements of the new philosophy onto existing production systems, and these nearly always fail. Methods such as single sourcing, JIT "total quality" and quality circles (see Chapter 10) initiatives are put in place, often with considerable publicity and exposure, and often with the support of outside consultants, only to fail after a few months or a few years. The piecemeal grafting of such ideas on to a production system that follows the principles of traditional mass production is almost guaranteed to fail or to have only limited short-term success, because the transplant is inherently unstable in such a host organization.

Experience and reason both reveal that transformation of a production system to the new philosophy cannot be imposed by statements of intent or by the piecemeal introduction of selected aspects. It can be achieved only by top-level commitment, based on a thorough understanding of the principles involved. This commitment and understanding must permeate all levels of management, so the introduction must be preceded by a fundamental review of functions and responsibilities and a systematic process of training. The principles of production and of the new philosophy must provide the basis for the necessary changes. Since the managers involved might have been trained and experienced in traditional methods it is essential that they fully absorb the new culture and discard management ideas based on "scientific" concepts and functional divisions.

The new principles can take hold only if all of the key people, i.e. the managers and the production workers, are trained to understand the methods and benefits. To create the necessary climate of receptiveness and team spirit the gap of knowledge and skills that traditionally exists between managers and production people must be reduced. Modern competitive production must not be seen as a low-skill task. People are skilled by nature, and management's task is to develop and exploit these skills to the benefit of the enterprise and of the individuals. Only then will the production people be empowered and motivated to improve quality and productivity and to apply the methods of the new philosophy.

Similar considerations apply to supplier involvement and JIT production. The suppliers must be selected and developed as genuine partners, with the same shared commitment to the enterprise.

The new philosophy cannot be imposed quickly on an existing organization, and it is counter-productive to demand early measurable "results" in terms of increased productivity or reduced costs. In fact, there will inevitably be disruptions to production and slow initial build-up afterwards. The training effort and the discussion and planning that will be necessary will cause further apparent delay and will be expensive. However, the process must not be driven by arbitrary schedules or productivity objectives. Instead, it must be driven by commitment to create the environment in which continuous improvement becomes self-sustaining. The self-sustaining improvements will then drive continual productivity growth, which will quickly exceed that which could be generated by traditional management action and will more than offset the initial effort and delay. It is not possible to "prove" this in advance, but it must be accepted as an act of faith by the people involved.

It is obviously much easier to adopt the new philosophy concurrently with the start of new product designs, as indeed was done in Japan. Since design, of the product and of the production processes, including the contributions of suppliers, is an essential feature of the whole approach, making the changes for an existing product design is likely to be much more difficult, and to generate less benefit. It is necessary to take account of the expected future market and the competition. However, there are not likely to be many production operations that justify retaining the traditional approach.

TECHNOLOGY IMPACT

Modern developments in technologies and in production methods strongly influence the ways in which the new production philosophy can be applied.

Almost all modern electronic products are manufactured on highly automated lines. Circuit board manufacture, component placement and attachment, and circuit test all require heavy investment in facilities, but these can then be used for manufacturing an almost unlimited range of product types. In many cases, only the final assembly and test stages are performed by people. As a result, many companies now outsource electronics manufacture to specialist businesses, as was discussed in Chapter 5.

For some high-volume non-electronic products, particularly cars, robotic systems perform a range of manufacturing tasks. However, these are mostly product-specific, so that most of this type of manufacture is performed in-house.

Manufacturing planning and control can be aided by the use of appropriate software. Manufacturing resource planning (MRP) software performs tasks such as scheduling, progress monitoring and reporting. The business software systems mentioned in Chapter 5 include MRP capabilities as well as supplier management, etc., generally within the ERP systems*.

PRODUCTION INSPECTION AND TESTING

Despite best efforts to control all of the processes of production, it is inevitable that some items will be made that are defective. Defects can be created at all stages of production, including the use of defective parts from suppliers. Defects may be created by processes or by people. Despite the fact that it is always more economical to prevent defects than to find and correct them, or to deliver defective products, it is necessary to consider how defects can be economically detected and corrected.

---

* The early MRP systems, such as MRP II, were based on the traditional approach to manufacturing planning. I presume that the new ERP software offerings deal with JIT.

Defects are detected by inspection and by testing. Inspection includes all visual checks specified in a production schedule, as well as manual or automatic dimensional or other checks, such as for colour or smoothness. Test includes all other measurements using instruments such as digital testers or dynamometers.

Inspection and test do not directly add value to the product. They impose additional costs, since they are processes that require resources and take time. Inspection and test equipment can be expensive and it uses factory space. To these costs must be added the management effort of selection, purchasing and training, and operating costs of manning, maintenance and calibration. In addition, inspection and test are never 100% effective: they usually fail to detect a proportion of defective items and they can falsely identify good items as defective.

Since all test and inspection adds to costs, without adding value to good product (unless, as is sometimes the case, a test is performed to tune or adjust), it should always be justified carefully and eliminated when practicable. The best way to achieve this is to work to prevent defects from being created in the production processes. This ideal can be approached by following the principles of design and of development testing described in the two previous chapters, and the principles of production, described earlier in this chapter. Teamwork, training, attention to detail and continuous improvement can lead to the cost-effective elimination of tests, particularly at intermediate stages of production.

Inspection and test costs can also be reduced by integrating them into the manufacturing processes. A simple and effective way to do this is to make inspection by production workers an integral part of their tasks, so that they inspect the quality of upstream processes before performing their manufacturing operations. This eliminates the need for separate inspectors and engenders teamwork. It also ensures that defects are detected as early as possible, so that immediate corrective action can be taken. This approach to inspection is intrinsic to the modern principles of production described earlier.

Automated inspection and test processes can also be integrated into the production processes to minimise delays and costs. In-process gauging on machine tools and on robotic welding and assembly lines are examples of this.

Since inspection and testing add costs, it might seem reasonable to inspect or test a sample of production items rather than every item in order to monitor and control quality. Statistical aspects of inspection and test are discussed in the next chapter.

Stress Testing

Testing of production items can sometimes be made more effective by using increased or "accelerated" stresses, as described for development testing. Accelerated testing, also called environmental stress screening (ESS), is appropriate whenever the increased stresses are more likely to cause defective items to fail than to damage good items. This is usually the case with electronic components and assemblies, and with other products that can include defects that are not otherwise easily detected. For example, it is standard practice to test pressure vessels at pressures higher than their maximum rated pressure (with liquid, to prevent explosion): this will cause vessels with material defects to fail but the single pressure application does not damage those that are defect free. Many electronic components, particularly integrated circuits, are also subjected to stress screening, or "burn-in", during which they are tested at high temperature for several hours (up to 168 hours used to be specified for "military specification" components, purely because that is the number of hours in a week!).

The nature and duration of the stresses to be applied, including where relevant combined stresses, must be determined for each product, taking account of experience, the nature of the processes, the number or proportion of defects expected, damage limits and the importance of detecting defects. As with other test decisions, this is often difficult, uncertain and changeable. For example, it has been normal practice to perform stress screening on electronic assemblies, involving several cycles of high and low temperature operation or thermal shock, to detect problems such as damaged components or faulty solder connections. However, recent developments in surface-mounted components, which use much smaller, more closely spaced solder connections that must withstand the mechanical stress reversals imposed by the temperature changes, has led to the need to review the way that this type of test is applied.

Testing in production is described in detail in [14].

CONCLUSIONS

The world's highest levels of manufacturing productivity are achieved by the companies that recognize that the traditional approaches to production management cannot create a self-sustaining impetus towards continuous improvement. They also recognize the need for all of the people and suppliers involved to be given the authority and responsibility for productivity enhancement, starting with the early designs and continuing into production. Finally, they recognize the essential positive connection between productivity and quality, as will be discussed further in the next chapter. The most productive companies realize that the improvement process is unending, and they demand continuous improvement from all of their people, processes and suppliers.

Continuous improvement of productivity is a deductive process, requiring careful analysis of the current situation, planning the improvements, trying them out and then acting on the results to ensure that the improvements are permanently installed. Deming called this the "Plan, Do, Check, Act" (PDCA) cycle. The many small improvements create a continuous forward flow. By contrast, attempts to maintain close control of a static situation, combined with occasional management-driven improvement "campaigns", seldom work as effectively, and are often counter-productive. The occasional failure in a continuous, step-by-step improvement programme is not a disaster, whereas campaigns that fail or have limited success are demoralizing and can lead to entrenchment and unwillingness to try further changes.

The deductive, continuous philosophy of improvement must rely upon the enlightened support of all of the people involved. Making this happen is the task of top management.

10

QUALITY, RELIABILITY AND SAFETY

If the people involved in design and production of a new product never made mistakes, and there was no variation of any feature (dimensions, parameters) or in the environment that the product would have to endure, then it would be relatively easy to design and manufacture products which would all be correct and which would not fail in use. Of course, engineering reality is seldom like this. Engineering designers must take account of variation, and also of the wear and degradation imposed by use and time. Production people must try to minimize the effects of process variations on quality, yield, reliability and costs. Maintenance people and users must work to keep the products serviceable.

The more that variation can be reduced and quality and reliability improved, the greater will be the benefits to the manufacturer and the user. Fewer failures during development will reduce development costs (redesign, re-test, delays, etc.). Less variation and fewer failures in production will reduce production costs (rework, scrap, work in progress, investigations, etc.). Finally, the reliability in service will be improved, resulting in enhanced reputation and thus increased sales, lower warranty costs, and greater profit opportunities. Throughout, the managers and engineers who would otherwise be busy dealing with failures would be freed to concentrate on new and improved products.

A very good example of the powerful impact of quality and reliability on competitive position is the ability of most Japanese car manufacturers to offer warranties extending to 3 years, in comparison with the one-year warranties offered by most of their Western competitors. Since the features that differentiate competing cars are otherwise very finely balanced, such a warranty is often a clear deciding factor. However, it can be offered only if the manufacturer is confident that the product will be very reliable.

THE COSTS OF QUALITY AND RELIABILITY

Failure Costs

There are two aspects to the quality and reliability cost picture. By far the largest, in nearly all cases, are the costs of failure. Failures generate costs during development, during production and in service. The further downstream in the process that causes of failures are discovered the greater is the cost impact, both in terms of the effect of the failure and to remove the cause, in nearly all cases.* Problems that cause failures or rejection after the product has been delivered add to the costs of warranty or service, and can also influence the product's reputation and sales. These internal and external costs of failures obviously depend upon the rates of occurrence.

Some of these costs, for example the cost of a re-design and re-test to correct a problem, the costs of production rework and scrap, and warranty repair costs are relatively easy to identify and quantify. These can be thought of as the direct failure costs. There are also indirect costs, for example management time involved in dealing with failures, staff morale, factory space needed for repairs, documentation, the extra test and measurement equipment needed for diagnosis and repair, the effects of delays in entering the market, delivery delays, and the effects on product reputation and therefore on future sales. Deming called these the "hidden factory", the cost of which usually exceeds the profit margin on the products concerned. In extreme cases failures can lead to product recalls, and litigation and damages if injury or death is caused. The indirect costs can be difficult to identify and quantify, but they are often much greater than the direct costs. It is important to note that failure costs begin soon after initial design and continue throughout the life of the product, over many accounting periods, and even beyond to future products. For a new product they are impossible to predict with confidence.

Obviously, we should have a strategy for minimizing these costs and risks. The strategy must be in place at the earliest stages of the

---

* The "Times Ten Rule" is often quoted: there will be a factor of 10 increase in costs for each further stage that a failure occurs. For example, a failure cause that is found during design might cost $100 to correct. The same failure found during development test might cost $1000 to correct, in production $10,000, and in service $100,000. Several cases show that this factor can be too low, with actual cost multipliers of 40 to 100 times being reported, and sometimes much higher if failures in service have serious consequences, such as product recall or causing injury or death.

project and it must extend through the life cycle. A sensible first step
is to make a determined attempt to forecast the likely range of failure
costs. Let us take a simple example.

	Totals $	Failure costs: Develop	Failure costs: Manufact. p.a.	Failure costs: Warranty p.a.	Failure costs: Reputation p.a.
Development cost	500000	40000			
Unit build cost	150				
Unit sales price	250				
Sales p.a.	5000				
Sales profit p.a.	500000				
Repair cost (factory)	50				
Failure costs (factory) p.a.(10%)	25000		25000		
Repair cost (field)	100				
Failure costs (field) p.a. (10%)	50000			50000	50000
Total failure cost p.a. (10%)	**75000**				
Fail cost p.a. % of profit (10%)	**15%**				
Total failure costs (5 year)	**665000**	40000	125000	250000	250000
Fail% of profit 5 years (10%)	**27%**				

Figure 10.1 Failure costs (10% fail)

A new product will cost $500,000 to develop, and $150 each to build. We expect to sell 5,000 per year, at $250 each. The average direct cost to repair a failure found during production is $50, and in the field is $100. If we assume that 10% will fail on factory test and in the field, we can easily derive the annual direct cost of failures. We can add the expected costs of failures during development and of the expected financial impact of the effect on reputation, as indirect costs. These assumptions are presented and analyzed in the spreadsheet (Figure 10.1).

Right away we can see how large a penalty is imposed by failures. The direct costs remove 15% of the annual profit on sales. If we look at the situation over the first 5 years of production, including the expected indirect costs (Figure 10.2) we see how the cumulative profit grows, with recovery of the investment in development in Year 3.

10% failure	Year 1	Year 2	Year 3	Year 4	Year 5	Year 6
SALES		1250000	1250000	1250000	1250000	1250000
CUM SALES		1250000	2500000	3750000	5000000	6250000
COSTS						
Development	500000					
Development failures	40000					
Production		750000	750000	750000	750000	750000
Factory failures		25000	25000	25000	25000	25000
Field failures		50000	50000	50000	50000	50000
Reputation		50000	50000	50000	50000	50000
Total Direct	540000	825000	825000	825000	825000	825000
Total direct (cum)	**540000**	**1365000**	**2190000**	**3015000**	**3840000**	**4665000**
Total	540000	875000	875000	875000	875000	875000
Total (cum)	540000	1415000	2290000	3165000	4040000	4915000
Profit	**-540000**	**-165000**	**210000**	**585000**	**960000**	**1335000**

Figure 10.2 Failure costs (10%), 6 years

We know that it would never be practicable to eliminate failure altogether, so it is not reasonable to expect that we could recover all of the failure costs. However, suppose that market research reveals that a similar competing product has a failure proportion of 5%, or we decide for any reason that this would be a realistic target. What would be the effect on costs?

	Totals $	Failure costs: Develop	Failure costs: Manufacture p.a.	Failure costs: Warranty p.a.	Failure costs: Reputation p.a.
Development cost	600000	60000			
Unit build cost	150				
Unit sales price	250				
Sales p.a.	6000				
Sales profit p.a.	600000				
Repair cost (factory)	50				
Failure costs (factory) p.a.(5%)	12500		12500		
Repair cost (field)	100				
Failure costs (field) p.a. (5%)	30000			30000	10000
Total failure cost p.a. (5%)	42500				
Fail cost p.a. % of profit (5%)	7%				
Total failure costs (5 year)	322500	60000	62500	150000	50000
Fail% of profit 5 years (5%)	11%				

Figure 10.3 Failure costs (5% fail)

First, we must decide how much more we might have to spend to improve quality, of the design and of production processes. Say that we decide to spend a further $100,000 on accelerated testing (Chapter 8), and that this will generate more failures during development, costing $60,000. We assume further that we will sell more (6000 instead of 5000 p.a.), since the higher reliability will attract more

5% failure	Year 1	Year 2	Year 3	Year 4	Year 5	Year 6
SALES		1500000	1500000	1500000	1500000	1500000
CUM SALES		1500000	3000000	4500000	6000000	7500000
COSTS						
Development	600000					
Development failures	60000					
Production		900000	900000	900000	900000	900000
Factory failures		12500	12500	12500	12500	12500
Field failures		30000	30000	30000	30000	30000
Reputation		10000	10000	10000	10000	10000
Total Direct	660000	942500	942500	942500	942500	942500
Total direct (cum)	**660000**	**1602500**	**2545000**	**3487500**	**4430000**	**5372500**
Total	660000	952500	952500	952500	952500	952500
Total (cum)	**660000**	**1612500**	**2565000**	**3517500**	**4470000**	**5422500**
Profit	**- 660000**	**-112500**	**435000**	**982500**	**1530000**	**2077500**

Figure 10.4 Failure costs (5%), 6 years

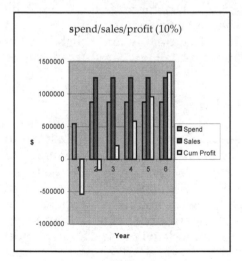
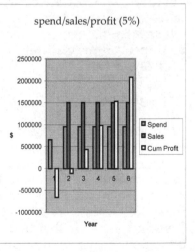

Figure 10.5 Cost comparisons

customers and make more items available for sale (items that fail in service must be replaced or repaired, so we must produce a quantity that cannot be sold; the lower the reliability the larger this stock must be). We also place a lower value ($1,000) on the annual reputation cost. Figure 10.3 shows the direct failure cost now falling to 7% of sales profit, and Figures 10.4 and 10.5 show a faster payback of the higher development cost and much higher sales profit over the five years in production ($2.1M compared with $1.3M). So the return on our extra development investment of $120,000 over 5 years will be about $800,000. Not a bad ROI.

Of course these figures can never be exact, nor can they all be stated with certainty. For example, the cost of the extra testing can be stated precisely, but how can we be sure that it will reduce the failure proportion to 5%? Also, they are based on projections into the future. Therefore they might not be convincing to financial people, particularly if their horizon is this year's figures.

We must overcome these problems by obtaining agreement on the input values and performing the analyses. Modern business software (Chapter 5), particularly life cycle cost programs, enable the analysis to be performed in much greater detail than the simple example above. Company and project management must take the long-term view. Finally, the expected benefits in terms of reduced failures must be taken on trust. They cannot be guaranteed or proven, but overwhelming evidence exists to show that quality and reliability can be improved dramatically with relatively small but carefully managed up-front effort.

Achievement Costs: "Optimum Quality"

In order to create reliable designs and products it is necessary to expend effort and resources. Better final designs are the result of greater care, more effort on training, use of better materials and processes, more effective testing, and use of the results to drive design improvement (product and processes). Better quality products are the result of greater care and skill, and more effective inspection and test. It seems plausible that there is an optimum level of effort that should be expended on quality and reliability, beyond which further effort would be counter-productive.

Ideally, we should seek to optimise the expenditure on preventing and detecting causes of failures, in relation to the costs that would arise otherwise. However, balancing the costs of achievement against the internal and external costs failures is difficult, since the balance depends upon a quantity, the rate or number of failures, that is highly uncertain and variable. It is also difficult to determine the true costs of many failures, and therefore these are often underestimated.

The conventional approach to this problem has been to apply the concept of the "optimum cost of quality". It is argued that, in order to prevent failures, the "prevention costs" would be increased, since more inspection and testing would be necessary. The "appraisal costs", which cover activities such as failure records and provision and calibration of test and measuring equipment, must also be increased in order to reduce failures. If these rising costs could be compared with the falling costs of reduced failures, an optimum point would be found, at which the total "cost of quality" would be minimized (see Figure 10.6).

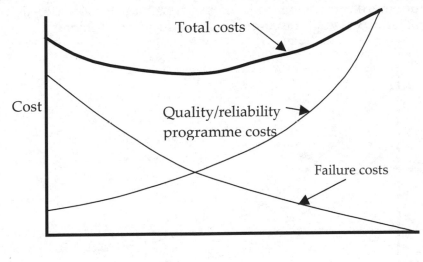

Figure 10.6 Quality and reliability costs: traditional view

The concept of the "optimum cost of quality" or "optimum quality level" at which this changeover occurs is embedded in much teaching and thinking about quality and reliability. Actually identifying the point in quantitative terms is not as easy, as discussed earlier. Nevertheless, the concept has often been used as a justification for sticking with the status quo.

It was Deming [5] who explained the fallacious nature of the "optimum cost of quality"*. The minimum total cost of failures occurs when they approach zero, not at some finite, determinable figure. The argument that achieving high quality necessarily entails high costs is dismissed by considering the causes of individual failures, rather than generalized factors such as failure rate. At any point on the curve of quality versus cost, certain failures are caused. Deming explained that, if action is taken to prevent the recurrence of these, there would nearly always be an overall saving in costs, not an increase. It is difficult to imagine a type of failure whose prevention will cost more than the consequences of doing nothing. Therefore the curve of cost versus failures moves downwards as quality is increased, not upwards (Figure 10.7). The idea that the curve must progress upwards was based on the "scientific" management theory, according to which higher quality could be achieved only by employing more skilled, and therefore more expensive labour, more inspectors, or more expensive material or processes (for example, higher precision machining).

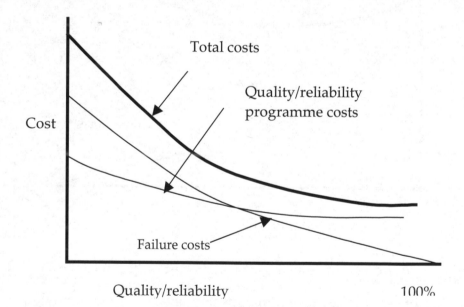

Figure 10.7 Quality and reliability costs: modern view

---

* Actually, Deming's arguments were presented in the context of production. This insight was probably the most important single idea that drove the Japanese post-war industrial revolution. He did not explicitly include the effects of reliability in service. Of course, if these are included the case is strengthened even more.

A further argument against the concept of an optimum quality level worse than perfection is the fact that actual determination of such a point is not practicable. The concept seems plausible, but any attempt to quantify the relationships fails, because of the great uncertainty involved. For example, how many failures, saving how much in costs, will be prevented by a particular test or inspection? The largest element of the costs of failures is usually the hardest to quantify.

Another major problem in this context is the fact that the development and production costs involved can be easily and accurately quantified, and they occur in the near future. However, as discussed above, the savings in failure costs are usually much more uncertain and they arise further ahead in time, often beyond the present financial plans and budgets.

The idea of an optimum quality somewhat lower than perfection can be a powerful deterrent to improvement. Since the "optimum" cannot be determined, it is tempting to believe that it is the status quo: what we are achieving now is as good as we can get. The concept has been a major inhibiting influence, and Deming's clear and simple exposure of its error was the key that unlocked the door to continuous improvement in quality and productivity.

The truth of the logic taught by Deming has been dramatically exposed by the companies that have taken it to heart. The original title of Deming's book was *Quality, Productivity and Competitive Position*, thus emphasizing the strong positive correlation of all three factors. This realization has been at the core of the success of the Japanese and other companies that have set new standards and expectations for quality and reliability, in markets such as cars, machine tools, electronics, and many others, while at the same time reducing their costs of development and production. The concepts follow inevitably from the principles of management described in Chapter 3, requiring a completely integrated team approach to the product, rather than the functional approach of "scientific" management. The talents and motivations of everyone on the team must be devoted to improving all aspects of quality. This approach to quality has been called "total quality management" (TQM). In this chapter we will consider how such excellence can be achieved.

VARIATION

All engineered products and systems must operate in conditions that vary. Diurnal and seasonal temperature variation, variation of mechanical stresses due to vibration or changing loads, and variation of electrical stress due to power and load fluctuations are a few examples of the many ways in which the environment and operating conditions can vary. In addition to these external variations there are many sources of variation internal to products. Any product that is manufactured in quantity will vary from item to item, due to the intrinsic variability of the processes by which they are made, and of the components of which they are comprised. For example, all machining operations include sources of variation that cause dimensional variation of parts, and casting and forging operations produce parts that vary in strength. Electronic component parameters are all variable, often with tolerances as wide as 10% or more. When components with variable dimensions, strengths and parameters are assembled into systems, then the systems will differ in performance.

The ways in which variations in external conditions and internal factors influence performance depends upon the ability of the design to withstand the variations that can be expected. In principle, if the designer is aware of the expected variability of operating conditions, dimensions and other parameters, he could analyze and test the design over the range of expected values. In practice, of course, he would identify the variables that knowledge and experience indicate are likely to be the most important ones, and he will evaluate the performance of the design at high and low values of these. For example, if the specification for an electronic filter circuit states the frequency response required, component parameter values can be selected to provide the specified performance. If there is an allowable tolerance on the frequency response, the filter performance can be calculated (or simulated) with the individual parameters at the high and low ends of their tolerances. If necessary, tighter tolerances would be chosen for parameters to which performance is shown to be sensitive, or some other form of compensation might be used such as a trimmer resistor, or circuits could be built and tested, and adjusted or repaired if they operate outside the specification. Clearly, neither of these would be optimal solutions to the problem: it would be far better if the design could be made to be sufficiently insensitive to the parameter variations that no compensation, adjustment, or selection was necessary.

This can of course be done. However, the designer must also take into account the fact that some of the circuit parameters are sensitive to external variation, particularly in this case to temperature. He must therefore include the effects of temperature extremes in his analyses.

Some variations are also time-dependent, so that over a period of operation, particularly at high stress, the parameter may drift permanently in one direction. The designer must be aware of this type of variation also, and must decide whether it is relevant to his application. If it is, it must also be included in the analysis.

If the designer calculates the effects of each variable on the operation of the circuit, one calculation for each variable at its high and low tolerance extreme, he will have to repeat the calculation 2^n times, where n is the number of variable parameters: for a simple high or low pass filter this might be 2^5, or 32 calculations. Depending on his knowledge and experience, he might decide that some of the variables are not critical, and that he could therefore ignore them and concentrate on those that are more important, in this case for example the slew rate of the op-amp.

In design situations such as this, in which the effects of all parameter values can be determined because the system can be fully described over its operating range by known equations, analyzing the effects of variation is straightforward. If, however, other variables exist that are not as well understood the problem becomes more difficult. For example, if the filter is supposed to operate at high frequencies, then other parameters, unknown to the designer, such as the mutual capacitance and inductance between conductors, or parasitic parameters within or between components, or even between the circuit and other circuits, might affect performance. These parameters might be fixed or variable. The designer is then faced with a much more difficult problem, and much more analysis is required if the circuit is to be designed to be insensitive to their variation.

A further problem arises if the effects of variables on performance are not independent, but interact. For example, the circuit's cut-off frequency might be affected more by the op-amp slew rate when the gain is high than when it is low. This effect is called a two-factor interaction. Higher order interactions can also exist, but they are rare in engineering. If the designer analyses the circuit at each combination of high and low tolerance values, he will not detect the effects of interactions such as these. Interactions between variables can be deduced from the equation of the system, but, as we will discuss later, not all designs can be easily and completely described mathematically.

Variation and Statistics

So far we have discussed variation in a purely deterministic context. However, if we buy resistors or machined parts, the spread of values between the tolerance limits is not uniform but is distributed so that most items will have values near the average and progressively fewer near the extremes. Therefore, to analyze designs for parameters at their tolerance limits can be considered to be unduly pessimistic. The method is sometimes called worst-case analysis (WCA). In fact, in the filter example the probability that any circuit will actually be built with all components at one extreme or the other of their tolerance ranges will be very remote. Instead, we can use statistical methods to analyze the circuits if we need to determine, for example, what proportion of manufactured circuits will be likely to perform outside the specification. In order to do this we will need to know the statistical distributions of all of the parameters. We will also need to know how to apply the appropriate statistical methods. As a first assumption we could decide that the parameters are distributed according to the "normal" distribution, which is the probability function that most closely describes the great majority of natural variation, such as people's heights and IQs and monthly rainfall. It also describes quite well many variables encountered in engineering, such as dimensions of machined parts, component parameter values, and times to failure due to material fatigue. The "normal" (or Gaussian) distribution is therefore widely used by statisticians and others and it is taught in all basic statistics courses and textbooks. Therefore it could be a reasonable starting point for a statistical analysis of the filter circuit.

We could assume that the nominal parameter values are the averages (or means) of their distributed values, and that, say, 95% of the parameters are within tolerance. (Alternatively, we could assume a standard deviation for the distribution: this is the measure of the spread of parameter values. We could say, for example, that the tolerance limits are 2 standard deviations from the mean, which is approximately the same as saying that 95% are within tolerance.) We could then simulate the manufacture of circuits with component parameters selected randomly (but normally distributed about the average), measure their performance, and determine the proportion defective. This is called a "Monte Carlo" analysis. The method is used for studies of logistics and processes, as well as for design tolerance analysis. It is practicable only when run on a computer, since the task of analyzing large numbers of possible situations, using randomly generated values for all the variables, is otherwise too tedious. Monte Carlo simulation facilities are built into most CAE software. It is a powerful addition to the design analysis capability since it enables designs to be evaluated for sensitivity to parameter variations and for estimating production test yields.

Monte Carlo analysis can be useful, but the method assumes that the behaviour of the system is fully and accurately modelled by the simulation, in this case the CAE software. However, many real engineering designs and processes cannot be described mathematically with such ease, completeness and accuracy. For example, a high-gain, low-noise amplifier might be very sensitive to parasitic component parameter values that are not closely controlled or specified, or the yield of a metal casting process might be affected by several variables in ways that are not predictable by any theory. If we are uncertain about the exact relationships between parameters and their effects, or we have no theoretical model, how can we analyse the effects?

Statistical Experiments

R.A. Fisher invented the technique known as statistical design of experiments (DOE) to analyse the effects of variable inputs on the yields of crops. Here the variables are different fertilisers, trace elements, soil conditions, etc. In this kind of situation there are no scientific theories on which mathematical models of cause and effect can be based The researchers know what the variables are, but they do not know how large their effects are, or whether the effects are interactive. They cannot therefore calculate the effects or simulate the system.

The researchers could perform a series of traditional scientific experiments, in which all variables are held constant (the "control" factors), except the one whose effect is being studied, and this is then repeated for other variables. Testing one variable at a time can be very time-consuming (particularly in agricultural work). Also, whilst it makes no assumptions about the cause-and-effect relationships, it cannot detect interactions.

In a statistically designed experiment, a range of tests is performed in which every variable is set at its expected high and low values (and possibly also at intermediate values), and the effects are recorded for each set of values. The results are analyzed using the technique of analysis of variance (ANOVA), which determines the magnitude of the effect of each variable, and of interactions between variables. The only assumptions necessary are that the variables are normally distributed and that their effects are approximately linear over the range tested.

Statistical experimentation is widely used in fields such as process optimization in the chemical and pharmaceutical industries, medical research and agriculture. It is also used in engineering, but not as widely as it should be. This is mainly because few engineers have been taught the method, but also because most engineering product and process designs can be analyzed and tested using traditional, deterministic, theory-based methods. The main applications have been for problem solving and for optimising production processes, such as machine soldering of electronic assemblies. However, since variation exists in products, processes and environments, product and process designs can be refined and made more robust, economical and reliable by appropriate use of statistical experiments.

Genichi Taguchi developed a framework for DOE adapted to the particular requirements of engineering design. Taguchi suggested that the design process consists of three phases: system design, parameter design and tolerance design. In the system design phase the basic concept is decided using theoretical knowledge and experience to calculate the basic parameters to provide the performance required. Parameter design involves refining the values so that the performance is optimized in relation to factors and variations that are not under the effective control of the designer, so that the design will be "robust" in relation to these.

Tolerance design is the final stage, in which the effects of random variation of manufacturing processes and environments are evaluated to determine whether the design of the product and of the production processes can be further optimized, particularly in relation to cost. Note that the design process is considered to explicitly include the design of the production methods and their control. Parameter and tolerance design are based on the statistical design of experiments.

Taguchi separates variables into two types. "Control" factors are those variables that can be practically and economically controlled, such as controllable dimensional or electrical parameters. "Noise" factors are the variables that are difficult or expensive to control in practice, although they can be controlled in an experiment, e.g. ambient temperature or parameter variation within a tolerance range. The objective is then to determine the combination of control factor settings (design and process variables) that will make the product have the maximum "robustness" to the expected variation in the noise factors. The measure of the robustness is the "signal-to-noise ratio", which is analogous to the term used in control engineering. The experimental framework is as described earlier. Taguchi argued that in most engineering situations interactions do not have significant effects, so that much reduced, and therefore more economical, "fractional" experiments can be applied. When necessary, subsidiary or confirmatory experiments can be run to ensure that this assumption is correct. Taguchi developed a range of such design matrices, or "orthogonal arrays", from which the appropriate one for a particular experiment can be selected.

The Taguchi approach has been criticized by some statisticians and others for not being statistically rigorous and for under-emphasizing the effects of interactions. Whilst there is some justification for these criticisms, it is important to appreciate that Taguchi has developed an operational method that deliberately economizes on the number of trials to be performed in order to reduce experiment costs. The planning must take account of the extent to which theoretical and other knowledge, for example experience, can be used to generate a more cost-effective experiment. For example, theory and experience can often indicate when interactions are unlikely or insignificant.

It is arguable that Taguchi's greatest contribution has been to foster a much wider awareness of the power of statistical experiments for product and process design optimization and problem solving. The other major benefit has been the fostering of the need for an integrated approach to the design of the product and of the production processes.

A statistical experiment can always, by its nature, produce results that seem to be in conflict with the physical or chemical basis of the situation. The probability of a result being statistically significant in relation to the experimental error is determined in the analysis, but we must always be on the lookout for the occurrence of chance results that do not fit our knowledge of the processes being studied. That is not to say that such results would be dismissed, only that we can legitimately use our engineering knowledge to help interpret the results of a statistical experiment. The right balance must always be struck between the statistical and engineering interpretations. If a result appears to be highly statistically significant then it is conversely highly unlikely that it is a perverse one. If the engineering interpretation clashes with the statistical result and the decision to be made based on the result is important then it is wise to repeat the experiment, varying the plan to emphasize the effects in question.

Statistical experimental methods of engineering design optimization can be effective and economic. They can provide higher levels of optimization and better understanding of the effects of variables than is possible with purely deterministic approaches when the effects are difficult to calculate or are caused by interactions. However, as with any statistical method, they do not by themselves explain why a result occurs. A scientific or engineering explanation must always be developed so that the effects can be understood and controlled.

It is essential that careful plans are made to ensure that the experiments will provide the answers required. This is particularly important for statistical experiments owing to the fact that several trials are involved in each experiment, and this can lead to high costs. Therefore a balance must be struck between the cost of the experiment and the value to be obtained, and care must be taken to select the experiment and parameter ranges that will give the most information The "brainstorming" approach is associated with the Taguchi method, but should be used in the planning of any engineering experiment. In this approach all involved in the design of the product and its production processes meet and suggest which are the likely important variables and plan the experimental framework. The team must consider all sources of variation and their likely ranges so that the most appropriate and cost-effective experiment is planned. A person who is skilled and experienced in the design and analysis of statistical experiments must be a team member, and may be the leader. It is important to create an atmosphere of trust and teamwork, and the whole team must agree with the plan once it is evolved.

Statistical experiments are equally effective for problem solving. In particular, the brainstorming approach often leads to the identification and solution of problems even before experiments are conducted.

Most of the available statistics software includes DOE capabilities. Statistical experiments can be conducted using CAE software, when the software includes the necessary facilities such as Monte Carlo simulation and statistical analysis routines. Of course there would be limitations in relation to the extent to which the software truly simulates the system and its responses to variation, but on the other hand experiments will be much less expensive, and quicker, than using hardware. Therefore initial optimization can often be usefully performed by simulation, with hardware experiments being run to confirm and refine the results.

Statistics and Engineering

Whilst statistical methods can be very powerful, economic, and effective in engineering applications, they must be used in the knowledge that variation in engineering is in important ways different from variation in most natural processes. The natural processes whose effects are manifested in, for example, people's heights, are numerous and complex. Many different sources of variation contribute to the overall height variation observed. When many distributed variables contribute to an overall effect, the overall effect tends to be normally distributed. If we know the parameters of the underlying distributions (means, standard deviations), and if there are no interactions or we know what the interaction effects are, we can calculate the parameters of the overall distribution. These properties are also true of engineering or other non-natural processes which are continuous and in control: that is, if they are subject only to random variation.

Natural variation rarely changes with time: the distributions of people's heights and life expectancies and of rainfall patterns are much the same today as they were years ago, and we can realistically assume that they will remain so for the foreseeable future. Therefore any statistical analysis of such phenomena can be used to forecast the future, as is done, for example, by insurance actuaries. However, these conditions often do not apply in engineering. For example:

- A component supplier might make a small change in a process, which results in a large change (better or worse) in reliability. Therefore past data cannot be used to forecast future reliability using purely statistical methods. The change might be deliberate or accidental, known or unknown.
- Components might be selected according to criteria such as dimensions or other measured parameters. This can invalidate the normal distribution assumption on which much of the statistical method is based. This might or might not be important in asses-sing the results.
- A process of parameter might vary in time, continuously or cyclically, so that statistics derived at one time might not be relevant at others.
- Variation is often deterministic by nature, for example spring deflection as a function of force, and it would not always be appropriate to apply statistical techniques to this sort of situation.
- Variation in engineering can arise from factors that defy mathematical treatment. For example, a thermostat might fail causing a process to vary in a different way to that determined by earlier measurements, or an operator or test technician might make a mistake.
- Variation can be non-linear, not only continuous. For example, a parameter such as a voltage level may vary over a range, but could also go to zero, or a system might enter a resonant condition.

These points highlight the fact that variation in engineering is caused to a large extent by people, as designers, makers, operators, and maintainers. The behaviour and performance of people is not as amenable to mathematical analysis and forecasting as is, say, the response of an engine to air inlet temperature or even weather patterns to ocean temperatures. Therefore the human element must always be considered, and statistical analysis must not be relied on without appropriate allowance being made for the effects of motivation, training and management, and the many other factors that can influence performance, cost, quality and reliability.

Finally, it is most important to bear in mind, in any application of statistical methods to problems in science and engineering, that ultimately all cause-and-effect relationships have explanations, in scientific theory, engineering design, process or human behaviour, etc. Statistical techniques can be useful in helping us to understand and control engineering situations. However, they do not by themselves provide explanations. We must always seek to understand causes of variation, since only then can we really be in control.

Process Variation

As stated earlier, all engineering processes create variation in their outputs. These variations are the results of all of the separate variables that influence the process, such as backlash and wear in machine tool gears and bearings, tool wear, material properties, vibration and temperatures. A simple electrical resistor trimming process will produce variation due to measurement inaccuracy, probe contact resistance variation, etc. In any manufacturing process it is obviously desirable to minimize variation of the product. There are two basic ways in which this can be achieved: we can operate the process, measure the output, and reject (or re-process if possible) all items that fall outside the allowed tolerance. Alternatively, we can reduce the variation of the process so that it produces only good products. In practice both methods are used, but it must be preferable to make only good products if possible. We will discuss the economic and management implications of these two approaches later, but let us assume that we will try to make all products within tolerance.

First of all, we must ensure that the process is inherently capable of maintaining the accuracy and precision required. "Accuracy" is the ability to keep the process on target: it determines the average, or mean, value of the output. "Precision" relates to the ability to make every product as close to target as possible: it determines the spread, or standard deviation, of the process. Having selected or designed a process that is capable of achieving the accuracy and precision required, we must then control it so that variation does not exceed the capability.

The "process capability index" (CPI) is defined as the tolerance divided by the interval of 3 times the process standard deviation. This can be adjusted for processes that are off-centre. A process capability index of 1, therefore, means that, on average, not more than 0.3% of products will be outside tolerance. This assumes of course that the output is normally distributed and that the process is under control. Both of these assumptions must be treated with caution.

The assumption of statistical normality in engineering applications was discussed earlier. No manufacturing process can be truly "normal" in this sense for the simple reason that there are always practical limits to the process. By contrast, the mathematical normal distribution function extends to plus and minus infinity, which is clearly absurd for any real process, even natural ones. Therefore it makes little sense to extrapolate far either way beyond the mean. Also, even if engineering processes seem to follow the normal distribution close to either side of the mean, say to plus or minus 1 or 2 standard deviations (90% to 95% of the population), the pattern of variation at the extremes does not usually follow the normal distribution, due to the effects of process limits, measurements, etc. There is little point, therefore, in applying statistical methods to analyze and forecast at these extremes.

A process is considered to be "in control" if the only variation is that which occurs randomly within the capability of the process. There must be no non-random trends or fluctuations, and the mean and spread must remain constant. If a process is capable, with, say, a process capability index of 4 or higher, and it is kept in control, in principle no out-of-tolerance products should result. This is the basic principle of "statistical process control" (SPC).

The principles of SPC were first explained by W.A. Shewhart in 1931. In his book, *The Economic Control of Quality in Production* [17], Shewhart explained the nature of variation in engineering processes and how it should be controlled. Shewhart divided variation into two categories. "Special cause" or "assignable" variation is any variation whose cause can be identified, and therefore reduced or eliminated. "Common cause" or "random" variation is that which remains after all special causes have been identified and removed: it is the economically irreducible variation left in the process.

The main method proposed by Shewhart for identifying the nature of variation in processes is the "process control chart". This is a simple plot of the mean values of successive samples of the production run, shown against the target for the process, drawn as a horizontal line on the graph. Ideally all of the plotted points should lie on the target line, but variations in the process will cause the plotted points to lie above or below this. Lines are also drawn parallel to the target line to indicate the limits of the process. Typically these represent 2 and 3 standard deviations of the process variation, which will have been determined by earlier measurements.

In addition to the chart of sample means, another chart, usually drawn just below the means chart, is used to plot the range of values within each sample. Limit lines are also drawn on the range chart.

The means chart (also called the X or X-bar chart) shows the accuracy of the process, and the range (R or R-bar) chart shows the precision. Figure 10.8 shows a typical control chart.

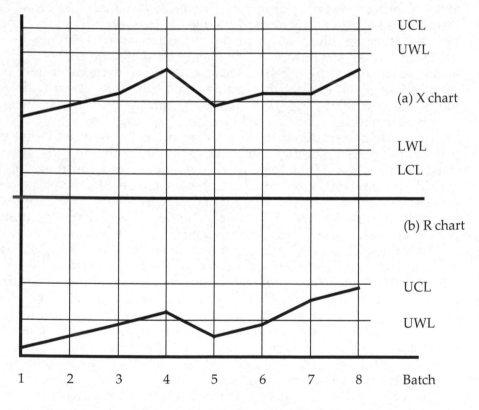

Figure 10.8 Process control charts.

The process is controlled by the operator plotting the mean and range values for successive samples of production batches on the control chart, and observing that they stay within the inner limit lines, called the upper and lower warning limits (UWL, LWL). If any point (mean or range) falls outside the warning limit, the process can be continued, but only if the next sample measurement falls within the warning limits. If successive measurements fall outside the warning limits, or any measurement falls outside the upper or lower control limits (UCL, LCL, or "action" limits), the process must be stopped and adjusted or corrected. The exact rules for particular processes may differ from those described, and there are many variations and extensions of the basic method related to production rates, batch quantities, sample sizes, derivation of limits and operating rules.

If the process is in control the lines will stay within the limits, apart maybe from occasional excursions beyond the warning limits. The fluctuations will represent the random variation in the process about the mean and within the process capability, and should therefore show no systematic trend or pattern. So long as this continues, no adjustments should be made because any change in the setting of the process mean to compensate for random fluctuation will increase, not decrease, the overall variation.

Control charts can be used to monitor any production process once the process is in control, and if the numbers being produced are large enough to justify using the method. They are simple to construct and to interpret. Many modern production tools, such as machining centres and gauges, include software that automatically generates control charts. They are one of the most important and effective tools for monitoring and improving production quality, and they should always be used, at the workplace, by the people running the process.

Shewhart pointed out the importance of control charts for process improvement. However, in much of industry they have been used mainly for monitoring, and not as part of an active improvement process. Also, an undue amount of attention has been given to the statistical aspects, so that books and teaching have tended to emphasize statistical refinement at the expense of practical utility, and the method is often perceived as being a specialists' tool rather than an aid to the process.

It was Deming, who had worked with Shewhart, who explained to Japanese industrialists the power of control charts for process improvement. This went together with his teaching that productivity and competitiveness are continuously enhanced as quality is improved, in contrast the traditional view that an "optimum" quality level existed beyond which further improvement was not cost effective. Later, Taguchi also took up this point in relation to design and used it as one of the justifications for the application of statistical experiments to optimize product and process designs. Statistical experiments, performed as part of an integrated approach to product and process design, can provide the most rational and most cost-effective basis for selecting initial control limits. In particular, the Taguchi method is compatible with modern concepts of statistical process control in production since it points the way to minimizing the variation of responses, rather than just optimizing the mean value. The explicit treatment of control and noise factors is an effective way of achieving this, and is a realistic approach for most engineering applications.

The control chart's use for process improvement is not based upon statistics. Instead, operators are taught to look for patterns that indicate "special causes" of variation. All process variation is caused by something, and the distinction between "common cause" and "special cause" lies only in the attitude to improvement. Any perceived trend or regular fluctuation can be further investigated to determine whether it represents a cause that can be eliminated or reduced. For example, Figure 10.8 shows two possible causes: the process mean seems to lie on average above the target, indicating that it should be reset. Also, the apparently regular fluctuation might be correlated with some other factor, such as room temperature or time of day, and the fact that the range fluctuations seem to move in step might provide a further clue.

Deming and Ishikawa taught Japanese production managers, supervisors, and workers how to interpret control charts and use them for process improvement. They also taught the use of other simple methods, such as Pareto charts and other graphical techniques, and Ishikawa developed the cause-and-effect diagram, an effective method for structuring and recording the efforts to determine causes of problems and variation. All of these methods (the "seven tools of quality") are used by the "quality circles", small groups of workers meeting during working time to determine and recommend ways of improving the processes they work on. Quality circles are the best known manifestation of Drucker's emphasis on people at the workplace being the most effective at generating improvements, as described in Chapter 3. The truth and effectiveness of these ideas have been dramatically demonstrated by many modern companies in highly competitive industries. Survival and growth in manufacturing industries depends as much on the fluent application of modern production quality management and methods as on product innovation and design. An excellent description of these methods is given in the book by David Hutchins, *In Pursuit of Quality* [18].

Statistical Sampling For Inspection And Test

Since inspection and testing add costs, it might seem reasonable to select a sample of production items, rather than every item, in order to monitor and control the level of quality. Of course we cannot detect all defects by sampling, but if the sample is carefully chosen to be representative of the production lot then it could provide information on the quality of the lot, in the same way as a public opinion poll is supposed to provide information on attitudes of the whole population. In some production situations it is not possible to test every item, if the test destroys or damages the item: munitions, fatigue test specimens, and electrostatic discharge tests on electronic components are examples; sample testing is necessary in such cases.

During the early 1940s statistical methods were developed for estimating the optimum number of a production lot that should be inspected or tested, as a randomly selected sample, in order to provide given levels of statistical confidence that the total lot contained not more than a given proportion defective. "Acceptance sampling" became a widespread practice, particularly for testing munitions. It was extended to other industries and processes such as electronic components, castings and many others. A typical requirement might state that a component must be produced to an "acceptable quality level" (AQL) of 0.1%, meaning that not more than 0.1% could be delivered defective, as determined by the statistical sample and the sampling rules specified. A later refinement was the "lot tolerance percent defective" (LPTD), used mainly for electronic components for which higher quality criteria were required and therefore larger samples had to be tested.

Deming drew attention to the basic flaw in the philosophy of acceptance sampling. Statistical acceptance sampling tables minimize the cost of sampling, not the cost of production. When the only option to no testing is to test a sample, as in a destructive test, sampling is the only way that assurance of a specified level of quality can be obtained by test. However, in nearly all other situations the true optimum decision for inspection or test is not how many to sample, but whether to perform it or not, i.e. zero or 100%. There is no theoretical optimum position between the extremes of zero or 100% test or inspection. We can estimate the decision point in terms of proportion defective if we know the costs of test and inspection and the costs of failures, but this point marks a discontinuity, and there is not a continuous relationship between the proportion sampled and the cost of production

For the great majority of modern products, the standards of quality required by the market are very high. For example, for most electronic components only a few per million of delivered items are likely to be defective. No statistical sample of a few tens or hundreds or even thousands can be sensitive enough to detect or measure such proportions. Therefore, as quality levels increase to match modern expectations acceptance sampling becomes progressively less relevant. The only way that such high levels of delivered quality can be achieved is for every defective item to be identified, and this cannot be achieved by testing only a sample. On the other hand, for some products manufactured using processes that are very well controlled, it might be economic to dispense with testing altogether. Obviously this is the optimum situation, which should be the objective in any production situation.

In practice, however, the analysis is not usually so straightforward. This is due to the large variability that often exists in quality. For example, components might be manufactured for a long period with very few defects, but a slight change in a process or material might cause a sudden deterioration. If it had been decided not to inspect or test the component this deterioration might cause expensive repairs later. On the other hand, it is not possible to forecast when or how often such problems might arise.

At the level of higher assemblies, such as an electronic system or an engine, the manufacturer must also decide on his test strategy. If he is receiving components that have been tested and only an extremely small proportion is expected to be defective, it is unlikely to be cost-effective to test them again before use. His component testing costs will be very high in relation to the subsequent saving in costs of repair. Therefore he should test none. However, the separate low probabilities of each component being defective combine by the product of the probabilities. Thus, for example, if an assembly uses 1000 components, each with a proportion defective of 10 per million, then the probability that the assembly will contain a defective component will be about 1%. In addition, the assembly processes can introduce further defects.

Therefore judgments on inspection and test must be made on the basis of experience and expectations, bearing in mind the uncertainty and variation of the main input to the analysis, the expected proportion defective. In most cases the decision is fairly clear-cut. For example, modern electronic components, even complex integrated circuits, are of such high quality that testing them before assembling them into circuits is definitely not cost effective, particularly as the cost of integrated circuit testers is very high. On the other hand, large electronic assemblies, consisting of many components and involving several manufacturing processes, should be subjected to 100% test. Therefore it is normal for such assemblies to be tested at the end of manufacture, and often also at earlier stages of assembly. Similarly, a mechanical product such as a diesel engine would be subjected to 100% testing, though the separate bearings and minor subassemblies such as oil pumps and injectors might not be.

Sampling inspection or test can sometimes be justified if the process being monitored is one which can suffer from large deviations from control which are not detected at the time. For example, the plastic moulding process used for encapsulating integrated circuits makes use of a mix of epoxy resin made up for a large number of components. If there is a problem with the process it is likely to affect many or all of the components that are encapsulated with that mix. Therefore, testing a small sample of each mix for adequacy of encapsulation is often a prudent policy.

In cases such as these there is no point in trying to derive statistically "optimum" samples and criteria. The test purpose is to detect when the process has moved out of control, but has not been detected at the time. No one can forecast the proportion that will be defective as a result, but the test should have a high probability of detecting gross changes. Therefore the number to be tested and the nature of the test must be based on practical knowledge, such as the cost of the test, the number that can conveniently be handled and the nature of possible problems.

RELIABILITY

If quality can be thought of as the excellence of a product at the time it is delivered to the customer, reliability is used in the engineering context to describe the ability of a product to work without failure during its expected time in use. A product's reliability therefore depends upon how well it is designed to withstand the conditions under which it will be used, the quality of manufacture, and, if appropriate, how well it is used and maintained.

Engineering products can fail in service for many reasons. These include:

- Variation of parameters and dimensions, leading to weakening, component mismatch, incorrect fits, vibration, etc. Design and manufacturing to minimize variation and its effects have been discussed earlier.

- Overstress, when an applied stress exceeds the strength of a component. Examples are mechanical overstress leading to fracture or bending of a beam or electrical overstress leading to local melting of an integrated circuit transistor or breakdown of the dielectric of a capacitor.

- Wearout, which is the result of time-dependent mechanisms such as material fatigue, wear, corrosion, insulation deterioration, etc., which progressively reduce the strength of the component so that it can no longer withstand the stress applied.

There are of course many other causes of failure, such as electromagnetic interference in electronic systems, backlash in mechanical drives, stiction and friction leading to incorrect operation of mechanisms, leaks, excessive vibration, and intermittent electrical connections. Failures are not always unambiguous, like a light bulb failure, but may be open to subjective interpretation, such as a noisy gearbox, a fluctuating pressure regulator or an incorrectly diagnosed symptom in an electronic system.

Failures can occur at any time in the life of a product. Some products can only fail once (light bulbs, microcircuits, rocket motor seals, geosynchronous satellites), since it is not economical or practicable to repair them. There may be different causes of failure, but the product's life is terminated whatever the cause. Other products can fail and be repaired a number of times (engines, TV sets, dishwashers, cars). The statistics of the times to failure of non-repairable products are similar to the actuarial statistics of human mortality. The statistics of failure of repairable systems are more like sickness statistics: people can have a headache, break a leg or develop cancer, and the consequences in terms of time off work, cost and danger are different.

Like human death and illness, failures of engineering products can occur in varying patterns through the life of the product. New products might fail soon after they are brought into use as a result of defects in manufacture. Wearout failure modes occur mainly in older products, such as bearings and drive belts that wear and cyclically stressed parts that suffer fatigue damage. Failures due to overstress can happen at any time in the life of the product, and their likelihood is not related to age.

Designers can in principle, and should in practice, ensure that their product designs will not fail under any expected conditions of variation, stress, wearout or for any other reason. The designers do not control variation in production, but, as explained earlier, they can ensure, together with the production people, that the effects of variations on performance and reliability are minimized and that appropriate tolerances and controls are designed into the production processes. Designers can prevent overstress failure if they understand the stresses that can be applied and ensure that adequate safety margins and protection are provided. They can protect against wearout failures by understanding the mechanisms and environments involved and by ensuring that the time to failure exceeds the expected life of the product by providing protection and, when appropriate, by designing a suitable inspection and maintenance plan Finally, they can protect against all of the other causes of failure by knowing how they occur and by attention to detail to prevent them.

Such care and attention in design, production and maintenance, far from being unrealistic and expensive, are in nearly all cases practicable and highly cost effective. Just as Deming pointed out the fallacy of an "optimum" level of production quality less than perfection, so design and development for any level of reliability less than 100% is wasteful and uncompetitive. Note that we are discussing the probability of no failures within the expected life and environment. The designer is not expected to cater for gross overstress or inadequate maintenance, though of course the design must include margins of safety related to the criticality of failure and the likely variations in stress and strength. The creation of a reliable design is nearly always more economical than creating a design that fails in service, as the example at the beginning of this chapter shows.

Reliability is a major determinant of a product's reputation and cost in service, and small differences in the reliability of competing products can greatly influence market share and profitability. If the reliability of a new product is perceived to be below expectations, or less than required by contract, serious losses or cancellation can result. Furthermore, it is usually extremely difficult and expensive to improve the reliability of a product after it has been delivered, particularly if the failures are due to design shortcomings.

Designing, developing, and manufacturing modern products and systems to be reliable is therefore crucially important. The principles of reliability engineering are, however, inherently simple. Good engineering leads to good products. The principles and methods described in this book are all essential aspects of engineering for reliability.

Reliability engineering methods and management are described in detail in O'Connor, *Practical Reliability Engineering* [19].

Quantifying Reliability

Since reliability is often expressed as a probability, the subject has attracted the attention of statisticians. Reliability can be expressed in other ways, for example as the mean time between failures (MTBF) for a repairable system, or mean time to failure (MTTF) for a non-repairable item, or the inverse of these, the failure rate or hazard rate. (Note that these measures imply that failures occur at a constant average rate: this is a convenient assumption that simplifies the mathematics, but which might bear little relation to reality).

Statistical methods for measuring, analyzing and predicting reliability have been developed and taught to the extent that many engineers view reliability engineering as being a specialist topic, based largely on statistics. This is also manifest in the fact that most books, articles, and conference papers on the subject relate to the statistical aspects and that nearly all university teaching of reliability is performed by mathematics, not by engineering faculties.

As we have discussed earlier in this chapter, the application of statistics to engineering is subject to practical aspects that seriously limit the extent to which statistical models can credibly represent the practical engineering situation, and particularly as a basis for forecasting the future. Since the cause of nearly every failure of a product in service is ultimately traceable to a person making a mistake, it is quite wrong and misleading to associate reliability with the design or the product, as though reliability were a physical characteristic like mass or power. The mass and power of an engineering product are determined by physical and chemical laws, which limit what a particular design can achieve. Every product built to that design would have the same mass and power, subject of course to small variations due to production variation. However, nature places no such constraints on reliability. We can make any product as reliable as we want or need to, and the only constraints are our knowledge, skill and effort. It therefore follows that a measurement of reliability is only a statement of history. If we record the number of failures of resistors in a large number of telephone switching circuits, and divide the number of failures by the total circuit operating time, we might say that we have measured the failure rate of resistors. However, any attempt to use this information to predict future reliability must be conditioned by the answers to questions like these:

- What were the causes of the failures? Were the causes poor quality of some resistors, poor assembly control, or design features that led to their being overstressed? (Note that if the first two causes predominate, repair should lead to fewer failures in future, but if the problem is mainly due to design, repair will not improve reliability).
- When the resistors failed and were replaced, were the new ones better than the ones that failed? Were the repairs performed correctly?
- Will future production of the same design use the same quality of resistor and of assembly? (Of course if failures have occurred action should have been taken to prevent recurrence).
- If the information on resistor reliability is to be used to predict the reliability of another system that uses similar resistors, do we know that the application will be identical in terms of stress, duty cycles, environment, test, etc. and, if not, do we know how to relate the old data to the new application?
- If the resistors that failed were mainly discrete metal film resistors, made by supplier X in 2002, and used in a voltage regulator circuit in which the tolerance was critical, can we apply the data to predict the reliability of a similar resistor made by supplier Y in 2004 to be used in circuit in which its precision is not critical?

These are not the only relevant questions, but they illustrate the problem of quantifying reliability at this level. In the great majority of cases the questions cannot be answered or the answers are negative. Yet we can confidently say that every 280KΩ 1% resistor will have a resistance of 280KΩ, plus or minus 2.8KΩ, and will handle up to its rated power in watts. The difference is that reliability measurements and predictions are based on perceptions, human performance and a huge range of variables, whilst parameter measurements and predictions are based on science.

To most engineers these comments might seem obvious and superfluous. However, measurements and predictions of reliability are made using just such approaches and the methods are described in standards and stipulated in contracts, particularly in fields such as military and telecommunication systems. For example, the U.S. Military Handbook for predicting the reliability of electronic systems (Military Handbook 217) provides detailed mathematical "models" for electronic component failure rates, so that one can "predict" the failure rate contribution per million hours of, say, a chip capacitor while being launched from a cannon, to an accuracy of 4 significant figures! Other organizations have published similar "data", and similar sources exist for non-electronic items. Reliability prediction "models" have even been proposed for software, for which there are no time-related phenomena that can cause failure. These methods are all in conflict with the fundamental principle that engineering must be based on logic (i.e. commonsense) and on science.

Statistical inference methods have also been applied to reliability "demonstration". The concept seems simple: test the product under representative conditions, for a suitable length of time, count the failures and calculate the reliability (e.g. MTBF). Then apply standard statistical tests to determine if the reliability demonstrated meets the requirement, to a specified level of statistical confidence. Note that statistical confidence has a specific mathematical meaning. For example, if Newton had observed 10 apples falling from a tree and had no other information or insight, he could not have stated with 100% statistical confidence that all apples will fall. In fact, his statistical confidence in such a statement would be zero, because there is an extremely large number of apples on trees and his sample was only 10. Of course his, and our, scientific confidence is 100%, even if we do not observe any apples falling, because we understand the effect of gravity.

Statistical "sequential" reliability demonstration tests make no allowance for this kind of reality. For example, if a product is tested and fails 10 times in 10 000 hours its demonstrated "best estimate" of MTBF would be 1000 hours. However, if 5 of the causes can be corrected, is the MTBF now 2000 hours? What about the failures that might occur in future, or to different units, that did not occur in the tests? Will the product be more or less reliable as it becomes older? Is it considered more reliable if it has few failures but the effects of the failures are very expensive or catastrophic?

The correct way to deal with failures is not merely to count them, but to determine the causes and correct them. The correct way to predict reliability is to decide what level is necessary and the extent of the commitment to achieving it. For example, a manufacturer of TV sets discovered that competitors' equivalent sets were about four times as reliable, measured as average repairs per warranty year. They realized that to compete they had to, at least, match the competitors' performance, so that became the reliability prediction for their new product. Note that the prediction is then top down, not from component level upwards. This is much more realistic, since it takes account of all possible causes of failure, not just failures of components. The prediction is top down also in the sense that it is management-driven, which is of course necessary because failures are generated primarily by people, not by components.

There are situations in which it is sensible to measure and analyze reliability and to make predictions by synthesis. Some examples will best illustrate these:

- Military aircraft electronic systems (avionics) have relatively high failure rates due to the very severe operating environments. The number of spares to be kept, the requirements for test and repair facilities and the logistics costs must be forecast and planned for. Therefore the reliability in service should be measured, though of course priority must also be given to improvement by identifying and correcting causes of failure. When predicting the reliability of a new avionic system, the similarities and differences must be considered, risks must be balanced against planned improvements and the contractual and management commitment must be determined. Note that if the prediction is based only on the raw failure data from the past, the predicted reliability will be no better than for the earlier systems, virtually denying the scope and motivation for improvement.
- If the new system being designed is comprised of subsystems and components for which good, credible reliability data exist, drawn from similar or identical applications, then this information can be used with care to synthesize the expected reliability of the new system. However, the potential for improvement must be studied and managed and taken into account in the prediction.
- Failure data can be subjected to statistical analysis to determine patterns and trends in support of engineering investigations when the causes of failure are uncertain.

These methods are described in [18].

SAFETY

Engineering products can cause hazards during operation or maintenance and if they fail. Well-publicized examples include the Challenger Space Shuttle explosion (failure of "O" ring seals on booster rocket casings), the Columbia loss (damage from separating polystyrene insulation), loss of aircraft (failures of engine mounting bolts, rupture of an engine turbine disc, electrical arcing of a fuel pump, etc.), and train derailments due to rails breaking. (It is interesting that, despite dire warnings from "experts", there have been no equivalent disasters caused by software). There have been many less dramatic incidents, resulting in death and injury to people at work, at home and while traveling. Safety incidents can obviously impact the reputation of a product and of the supplier. They can also generate very high costs, in product recalls, re-design and re-test. More significantly, they can lead to litigation and very high financial awards.

Design, development, manufacture and maintenance of engineering products must obviously seek to minimize the possibility of hazards. The methods that can be applied (primarily design analysis, test; see [18]) are mostly the same as for assuring reliability, so these should be extended as appropriate.

Increasingly, engineers are being required to "prove" the safety of new products and systems, particularly in applications such as public transport and large installations that present potential hazards, e.g. chemical plant and nuclear power stations. These formal, detailed statements are sometimes called "safety cases". The proof must be based on analysis of all potential modes of failure, including human failure, and of their consequences. The techniques used include failure modes, effects and criticality analysis and fault tree analysis, as used for reliability analysis and described in Chapter 7, as well as other techniques. The objective is to show that the probability of a single event or of a combination of multiple events that could cause defined hazards is acceptably low. The criteria for acceptability are usually laid down by the regulating authorities, and are typically for a probability not exceeding 10^{-8} per year for an accident causing loss of life. In order to "prove" such probabilities it is necessary to know or to agree what figures should be applied to all of the contributing events. We have discussed the incredibility of predicting reliability. Predicting hazard probabilities of this order is of course quite unrealistic. Any data on past events are likely to be only of historic significance, since action will almost certainly have been taken to prevent recurrence. Applying such probabilities to human actions or errors is similarly of extremely doubtful credibility. Also, accidents, particularly major ones, are usually the result of unforeseen events not considered in the hazard analyses.

It is of course essential that the hazard potentials of such systems are fully analyzed and minimized. However, it is equally important to apply commonsense to reduce the tendency to over-complicate the analysis. There is no value to be gained by attempting to quantify an analysis beyond the precision and credibility of the inputs. If the expected probability range of expected events is considered to be known to within an order of magnitude, it is absurd to present analyses that show the combined probability to a precision of several significant figures. It is also absurd to perform highly complex analyses when the causes and consequences can be sufficiently accurately estimated by much simpler methods. Such analyses can generate misguided trust in their thoroughness and accuracy, when in fact their complexity and implied precision can result in oversights and make them difficult to interpret or query. The KISS principle (*"keep it simple, stupid'*) applies to safety analysis just as much as it does to design.

The simplicity principle can be applied to quantification of risks by using easily understandable risk categories rather than pretending to know actual probabilities. For example, risks can be categorized on a range from "'very likely" to "extremely unlikely", or even "impossible", and these categories can be allocated rough probability ranges or figures of merit such as 5 for very likely to 0 for impossible. Any analysis performed using such figures is simpler to understand and to query and is more credible than analyses using pseudo "data".

Societal aspects of safety are discussed in Chapter 12.

QUALITY, RELIABILITY AND SAFETY STANDARDS

Quality, reliability and safety standards, and their limitations, are described in detail in [14], [17] and [18] and they are listed in the books' homepages. The most important are described briefly in the sections below.

Quality: ISO9000

The international standard for quality systems, IS09000, has been developed to provide a framework for assessing the extent to which an organization (a company, business unit or provider of goods or services) meets criteria related to the system for assuring quality of the goods or services provided. The concept was developed from the US. Military Standard for quality, MIL-Q-9858, which was introduced in the 1950s as a means of assuring the quality of products built for the U.S. military services.

The standard forms part of a series: IS09001 is relevant to organizations that conduct product design and development as well as production. ISO9002 relates to organizations that are producers only, and other parts of the series relate to the provision of services and software. Most industrial nations have adopted ISO9000 in place of their previous quality standards.

The original aim of supplier certification was to provide assurance that the suppliers of equipment operated documented systems, and maintained and complied with written procedures for aspects such as fault detection and correction, calibration, control of subcontractors and segregation of defective items. They had to maintain a "Quality Manual", to describe the organization and responsibilities for quality. It is relatively easy to appreciate the motivation of large government procurement agencies to impose such standards on their suppliers. However, the approach has not been effective, despite the very high costs involved.

The major difference between the ISO standards and their defence-related predecessors is not in their content, but in the way that they are applied. The suppliers of defence equipment were assessed against the standards by their customers, and successful assessment was necessary in order for a company to be entitled to be considered for contracts. By contrast, the ISO9000 approach relies on "third-party" assessment: certain organizations are "accredited" by their national quality accreditation body, entitling them to assess companies and other organizations and to issue registration certificates. The justification given for third-party assessment is that it removes the need for every customer to perform his own assessment of suppliers. A supplier's registration indicates to all his customers that his quality system complies with the standard, and he is relieved of the burden of being subjected to separate assessments ("audits") by all of his customers, who might furthermore have varying requirements. To an increasing extent, purchasing organizations such as companies, government bodies and national and local government agencies are demanding that their suppliers must be registered*. Many organizations perceive the need to obtain registration in order to comply with these requirements when stipulated by their customers. They also perceive that registration will be helpful in presenting a quality image and in improving their quality systems.

ISO9000 does not specifically address the quality of products and services. It describes, in very general and rather vague terms, the

---

* The European Community "CE Mark" regulations encourage registration. However, it is not true, as is sometimes claimed, that having ISO9000 registration is a necessary condition for affixing a CE Mark.

"system" that should be in place to assure quality. In principle, there is nothing in the standard to prevent an organization from producing poor quality goods or services, so long as procedures are followed and problems are documented. Obviously an organization with an effective quality system would normally be more likely to take corrective action and improve processes and service than would one that is disorganized. However, the fact of registration cannot be taken as assurance of quality. It is often stated that registered organizations can, and sometimes do, produce "well-documented rubbish". An alarming number of purchasing and quality managers, in industry and in the public sector, seem to be unaware of this fundamental limitation of the standards.

The effort and expense that must be expended to obtain and maintain registration tend to engender the attitude that the optimal standards of quality have been achieved. The publicity that typically goes with initial registration supports this. The objectives of the organization, and particularly of the staff directly involved in registration, are directed at the maintenance of procedures and audits to ensure that people work to them. It becomes more important to work to procedures than to develop better ways of doing things. Third-party assessment is at the heart of the ISO9000 approach, but the total quality philosophy demands close partnership between the purchaser and his suppliers. A matter as essential as quality cannot be safely left to be assessed by third parties, who are unlikely to have the appropriate specialist knowledge and who cannot be members of the joint supplier-purchaser team.

Defenders of ISO9000 say that the total quality approach is too severe for most organizations, and that ISO9000 can provide a "foundation" for a total quality effort. However, the foremost teachers of modern quality management all argue against this view. They point out that any organization can adopt the total quality philosophy, that it will lead to far greater benefits than registration to the standards and at much lower costs. The ISO9000 approach, and the whole system of accreditation, assessment and registration, together with the attendant bureaucracy and growth of a sub-industry of consultants and others who live parasitically on the system, is fundamentally at variance with the principles of modern management described in Chapter 4. It shows how easily the discredited "scientific" approach to management can be re-asserted by people and organizations with inappropriate motivation and understanding, especially when vested interests are involved.

ISO9000 has always been controversial, generating heated arguments in quality management circles. In an effort to cater for much of the criticism, ISO9000:2000 was issued. However, whilst this mitigates some of the weaknesses of the earlier version (for example, it includes a requirement for improvements to be pursued), the fundamental problems remain. Special versions of the standard have been developed by some industry sectors, notably automotive (ISO/TS16949:2002, replacing QS9000), commercial aviation (AS9000) and telecommunications (TL9000).

It is notable that the ISO9000 approach is very little used in Japan or by many of the best performing engineering companies elsewhere in the world, all of whom set far higher standards, related to the actual quality of the products and services provided and to continual improvement. They do not rely on "third-party" assessment of suppliers.

The correct response to ISO9000 and related industry standards is to ignore them, either as the basis for internal quality management or for assessing or selecting suppliers, unless they are mandated by customers whose importance justifies the expense and management distraction involved. If registration is considered to be necessary it is important that a total quality approach is put in place first. Compliance with the ISO9000 requirements will then be straightforward and the tendency to consider achievement of registration as the final goal will be avoided.

Reliability

U.S. Military Standard 785 is the original standard on reliability programmes. It described the tasks that should be performed and the management of reliability programmes. It referred to several other military standards that cover, for example, reliability prediction, reliability testing, etc.[*] U.K. Defence Standards 00-40 and 00-41 are similar to MIL-STD-785, but include details of methods. Non-military standards for reliability include British Standard BS5760 and the range of international standards in the ISO603000 family. Whilst these do include varying amounts of practical guidance, much of the material over-emphasises quantitative aspects such as reliability prediction and demonstration and "systems" approaches similar to those of ISO9000.

Safety Regulations and Standards

International, national and industry regulations and standards have

---

[*] The US DoD withdrew nearly all of these standards in 1995.

been created for general and for specific aspects of safety of engineering products. Managers must be aware of what regulations and standards are applicable to the projects for which they carry responsibility and they must ensure compliance. For example, the European CE Mark Directive is primarily related to safety, medical equipment must comply with US FDA regulations, and there are strict regulations for aviation equipment, high voltage electrical equipment, etc.

A recent development has been the "safety case", which is a document that must be prepared by the supplier and accepted by the customer. The safety case describes the hazards that might be presented and the ways by which they will be avoided or mitigated. The approach is applied in fields such as rail, power, process plant, etc., particularly when government approval is required. The safety case approach tends to be bureaucratic and "systems" based, rather like the ISO9000 approach to quality, and its effect on the safety of the UK railway system has not been encouraging.

An important new standard as far as engineering management is concerned is ISO/IEC61508, which is concerned with the safety of systems that include electronics and software. Nowadays, not many do not. The standard is without any practical value or merit. The methods described are inconsistent with accepted industry practices, and many of them are known only to specialist academics, presumably including the members of the drafting committee. The issuing of the standard is leading to a growth of bureaucracy, auditors and consultants, and increased costs. It is unlikely to generate any improvements in safety, for the same reasons that ISO9000 does not improve quality. Nevertheless, managers need to be aware of its applicability and how best to deal with it. It must not be ignored.

CONTRACTS FOR QUALITY AND RELIABILITY

Contracts for engineering projects placed by purchasers such as government agencies and other buyers of major equipment often include requirements for quality and reliability, including quantitative requirements, programmes of work and the application of quality and reliability standards. Obviously when these requirements are part of a contract they must be complied with, and it is then a major part of the project manager's task to ensure that the customer is satisfied in these respects. However, it is important to identify which tasks actually improve quality and reliability, and which are mainly or entirely concerned with meeting the requirements for documentation and reporting. The latter tasks, including for example formal plans and reliability predictions, can be delegated, as they do not present risks and they do not strongly influence or interfere with other project activities. However, the actual tasks of achievement and improvement, such as design analysis, test, training, organization and production quality assurance influence all aspects of the project and must therefore be managed as such.

It must not be assumed that compliance with the contract requirements for project activities and with quality and reliability standards will by themselves assure achievement of high quality and reliability. It is possible to comply with many of the contractual and standard quality and reliability requirements yet generate poor quality, unreliable products, as several major purchasing agencies have found to their cost. In particular, the international standard for quality systems, ISO9000, indicates no more than the fact that a "third party" accreditation agency is satisfied with the quality management system in use. The actual quality of the products or services provided is not assessed. As discussed above, the ISO9000 concept is not compatible with the modern approaches to management and quality. Project managers should not be deluded into thinking that it assures quality, either within their organization or from their suppliers.

To be as effective as possible, contracts for quality and reliability must comply with the principles of modern management of projects and production described in earlier chapters. Practical methods to ensure the right motivation include setting up partnership arrangements, coupled with warranties. The partnership approach ensures that problems are anticipated, identified and dealt with as quickly and effectively as possible. The warranty ensures that the supplier's top management remains committed to quality and reliability throughout the project, including its time in use.

A particular form of warranty that has proved to be extremely effective, benefiting both purchaser and supplier, is the reliability improvement warranty (RIW). In this form of contract the supplier guarantees to repair all failures and to supply serviceable spares, over a period of typically four years, for a fixed agreed cost. Other details might be included, such as a maximum repair time and a minimum reliability value to be achieved. In this form of contract the supplier is strongly motivated to develop and provide reliable equipment and the customer is assured of reliability and availability at a fixed cost. Obviously the RIW contract cost must be carefully negotiated in relation to the risks and benefits to both parties. RIW contracts are common practice for systems such as commercial aircraft and military systems.

There has been a trend in some recent major projects for the supply contract to include the provision of all support services (maintenance, spares, etc.) for a stated period. Such a contract places considerable motivation on the supplier to provide reliable equipment, since he bears most of the costs of failure. In effect, support is outsourced. We will discuss this approach in the next chapter.

MANAGING QUALITY, RELIABILITY AND SAFETY

Total Quality Management

Engineering project managers must take the lead on quality, reliability and safety, since all aspects of design, development, production and support are links that determine the levels achieved. Quality and reliability are critical contributors to development time and costs, production costs and project success. Safety hazards can present very high business risks. By delegating responsibility for these aspects the project manager hands over control of some of the most significant determinants of success or failure. He must therefore manage quality, reliability and safety, at the same time making the best use of specialists assigned to the project. The project manager must understand the full effects, in particular the relationships to competitiveness and costs. He must ensure that all engineers on the project are dedicated to excellence, and that they are trained and supported so that failures and hazards are avoided whenever practicable and corrected whenever found.

All engineers are quality, reliability and safety engineers. However, not all are trained and experienced in design analysis methods like failure modes and effects analysis, the relevant statistical and other analysis techniques, test methods, etc. Therefore some specialization is often appropriate. A small specialist group can also provide the focus for development of methods, internal support and consultancy and training. However, it is essential that the engineers performing quality, reliability and safety work on projects are integrated into the project teams, just like all the other contributors. It is unfortunate that, partly because of the perception that quality and reliability use statistical methods that most engineers find unfamiliar, and partly because many of the people (engineers and statisticians) engaged in this work have exaggerated the statistical aspects, quality and reliability effort is often sidelined and given low priority.

Depending upon the type of project, the hazard risks involved and the contract or regulatory requirements (discussed further in Chapter 12), safety aspects could be managed separately from quality and reliability. However, there should be close collaboration, since many of the analysis and test methods are complementary.

Management of the quality and reliability function should be combined to ensure that the product benefits from an integrated approach to all of the factors discussed earlier. The combination of quality and reliability responsibilities should be applied centrally, as well as on projects. However, some companies separate the roles of quality and reliability. They consider reliability to be related to design and development, and quality to production. This separation of functions can be justified in organizations and in projects in which design and development work predominates, or when production is undertaken elsewhere, for example by "outsource" subcontractors. However, it is nearly always preferable to combine the functions. A combined approach can be a powerful glue to encourage cooperation between design and production engineers, whereas separation can foster uncoordinated attitudes and approaches. A combined approach can also foster an integrated approach to training, of both quality and reliability specialists and of other engineers in quality and reliability topics. Since quality of design and of production are so integrally related to productivity, costs and reliability, many of the world's most competitive engineering companies combine the functions, and many use the term "quality" to encompass the integrated function. Sometimes the expression "off-line quality" is used to describe all of the work related to quality and reliability before production starts, and "on-line quality" to refer to related work after the start of production. This top-down, integrated approach to managing quality has been called total quality management (TQM).

"Quality System" Management

It is surprising in how many companies the "quality manager" is responsible primarily for the "quality system": procedures, audits and registration in accordance with standards. He often has a large staff, and is therefore a senior person. By contrast, reliability engineering support is provided by a few specialists, often with inappropriate qualifications or experience and very little authority or influence. Such an approach can easily comply with standards such as ISO9000, but it is guaranteed to create products that fail to achieve the highest levels of quality and reliability, and to be financially wasteful.

Obviously, it is essential that the "systems" aspects are effectively managed. For example, if the company has decided that ISO9000 registration is necessary, then the necessary procedures must exist, people must work to them and the internal and external audits must be organized. Someone must manage all of this. An appropriate solution is for a manager to be appointed, reporting to the quality manager, to look after all standards compliance, including quality standards. It must be clear that he is not responsible for the achievement of quality, reliability or safety: that is the project managers' responsibility. He should ensure that project staff are informed of the standards that apply, and he should manage the "systems" aspects with minimum interference with project work and with minimum bureaucracy and cost. Such a position would require a person capable of understanding both "system" and project priorities and of effectively dealing with the politics of external audits. This is not a job for a mediocre performer; the person appointed should be a competent engineering manager, and he should be judged for promotion on his performance as described, not left in a managerial backwater of low esteem.

Quality Circles

Quality circles are small groups of workers who meet regularly, typically weekly, to identify quality problems and implement improvements in the processes that they operate. They are given the authority to take action when they can, or to make suggestions when they cannot, for example to change a supplier or buy a new machine. A circle is led by one of the workers in the team or by their supervisor. The members are trained in the use of a range of techniques for analyzing data (the "seven tools of quality"), including simple statistical methods and SPC. They decide what to analyse and how to do it. Management's role is to provide the training, facilities, time and motivation.

The quality circles approach to quality improvement (*kaizen*) was at the heart of the Japanese post-war industrial revolution, and it is an integral and essential component of TQM. It was introduced in the 1950's as a result of the teaching of Deming, Juran and Ishikawa, and it spread rapidly throughout Japanese industry. Later, primarily due to Deming's work in the USA, it was widely adopted in other countries, particularly by the industries that were most affected by Japanese competition.

Note that the approach is consistent with Drucker's teaching that the people at the workface are the ones best qualified to manage their work and to generate improvements. It is described in [15], [17] and [18].

"Six Sigma"

The "six sigma" approach was originally developed in the USA by the Motorola company. It has spread to many other companies, mainly in the USA, particularly after its much-publicized application by GE under Jack Welch. It is based on the idea that if a process that is variable can be controlled so that all of its output that is within plus or minus six standard deviations of the statistical distribution will be within specification, then only about one in a million times will it produce a "defective" output. This assumes that the output is statistically "normally" distributed, as discussed earlier; this is of course, highly unlikely, especially at such extremes. The approach is supported by the use of statistical analysis tools to identify causes of variation and to implement improvements.

The main differences between six sigma and the quality circles approach is that six sigma is run by specialists, not by the people running the processes. Some of the analytical methods used are more advanced, including ANOVA and Taguchi. The trained six sigma people are given titles like "black belts", and it is their job to find problems and generate solutions. The whole operation is driven from the top, and is directed at achieving stated targets for measurable cost savings. External consultants are often involved in training and in execution.

Six sigma has been credited with generating significant improvements and savings. However, it is expensive. The management approach is "scientific", so it is arguable that the quality circles approach is a more effective philosophy.

CONCLUSIONS

Quality, reliability and safety have generated more misguided nonsense than any other aspect of engineering management. Inappropriate use of statistics and quantification, regressive standards, over-emphasis on "systems" and a huge growth industry of auditors and consultants have led to controversy and ineffective management. The specialist professions have reflected this: there has been continual argument in their professional journals, the most respected leaders such as Deming, Ishikawa, Hutchins and others have disassociated themselves from the trends, and the highest achievements are made by companies that stick to the methods recommended in this chapter. The journals of the quality profession in the USA and the UK have become almost totally devoted to ISO9000 and the "systems" idea of quality. Nearly all reliability and safety education and literature is mathematical and abstruse, and the published standards reflect this.

Arguments sometimes arise about what are the "best" methods for managing quality. For example:

- Q: Is six sigma better than TQM? A: No single set of methods, particularly if they are based on "scientific" management thinking, can be better than TQM. TQM is total.

- Q: Is TQM out of date? A: TQM can never be out of date.

- Q: Is ISO9000 consistent with TQM? A: No.

- Q: Can quality circles work in Western societies? A: Yes, of course, so long as managers understand and apply the philosophy correctly.

Achieving high quality, reliability and safety is primarily a deductive process. They are the results of attention to detail in design, development and production. Inductive, innovative thinking can create ideas for product and process design and often for problem solving. However, modern engineering products are combinations of large numbers of separate ideas, details, components, materials and processes, all of which must be correct if the product is to be economic, reliable and safe. Furthermore, taking every practical step to ensure that they are all correct is the optimum approach, as Deming taught. Such a disciplined, deductive, team-oriented approach often seems counter-intuitive, particularly to engineers working on projects that involve innovation. However, it is essential to provide the training, time and leadership necessary to ensure that every detail is considered and optimized.

Finally, it must be emphasized: well-managed effort and expenditure on quality, reliability and safety will always prove to be an excellent investment. Reference [17] provides an excellent overview for managers.

11

SELLING, USING AND SUPPORTING ENGINEERING PRODUCTS

SELLING

Most engineering products must be sold, and project success is then ultimately measured by sales and profits. There are two basic types of customer for engineering products: engineers and non-engineers. There are also overlaps, of course. The approach to selling (marketing, forecasting, advertising, support, etc.) must be aimed at the target customers.

There is little point in trying to impress non-technical customers with technical information. The selling points for the product must be the performance and other features that are important to the customer, and that he can understand. There are exceptions, for example when an innovation provides a major market advantage, or when it is considered that particular technical features will be likely to differentiate the product in the customer's eyes. However, this must be done carefully: customers will ultimately be impressed by technology only if it works in meeting their requirements for performance and reliability. Engineers on the project should curb their enthusiasm regarding techniques and innovation and should allow the marketing people to take the lead on sales strategy and tactics, though of course there must be close liaison.

Selling to engineers is different. Engineers as customers are obviously concerned with performance, price and service, but they are also interested in the technical aspects. Therefore full technical and performance information should be provided, and the people facing the customers must be able to provide this and to answer the questions that customers will ask. The sales people therefore must understand the product and its application. They should also know the capabilities and limitations in relation to competing products. Information on strengths should be used to sell the product. Information on limitations should be transmitted back to the project team so that they can consider how the product can be improved.

For all types of customer, price and service are of course also selling points that can be decisive, particularly when there are no other features that significantly differentiate the product from the competition. Design, development and production to minimize costs have been discussed earlier; however, it is worth stressing again the close and positive correlation between effort to improve quality and reliability and consequential reductions in development and production costs.

USE AND MAINTENANCE

The use of engineering products extends over time and in environments that impose stress and wear. Usage cannot improve a product: at best the product might be largely unaffected by use, but most products suffer some degradation over time. The way that the product is packaged, transported, stored and used, including the operating environment, care in handling and operating and the amount of use within a given time period can all affect degradation. The ways in which the designers should work to minimize the effects of these were discussed in Chapter 7. There are other ways by which degradation can be reduced.

For non-technical users the product must be designed so that misuse is very unlikely and correct use is intuitive. In particular, it must be unlikely that users can damage the product during normal use. User instructions can be provided which should advise on care and maintenance. However, for many products it cannot be assumed that users will study these. Even for quite complex modern products such as cars and video recorders, today's largely non-technical consumer expects to be able to use them without having to study manuals and undertake maintenance. This tendency is increased by the level of complexity of such products: whereas almost any motorist of 50 years ago could understand the workings of his car and could perform basic maintenance and troubleshooting, today's products require specialist skills and equipment to maintain them.

Engineers can be expected to be more understanding of the workings of the products they use professionally. For example, they should understand the need for lubrication of bearings and calibration of instruments, and they will be able to understand the technical aspects of user manuals. Therefore for products used by engineers full information should be provided in user manuals or data sheets. On-call or on-line technical assistance and advice is also important, and this should be provided as part of the support service. This should also be used as a source of information to influence design improvements and development of new products.

Maintenance Planning

There are two types of maintenance that engineering products must undergo. Preventive maintenance comprises the actions necessary to keep the product in good condition and to prevent failures. It includes cleaning, inspecting, lubricating and adjusting, as well as the replacement of parts that are known to wear out, or are seen to be wearing out, before they fail. Corrective maintenance is the repair of failures. Preventive maintenance can be planned. Corrective maintenance cannot be planned, though there can be flexibility concerning when repairs may be carried out if the failures do not prevent the product from being used.

Cars provide examples of all aspects of maintenance. The preventive maintenance includes cleaning the engine air filter, checking for tyre and brake pad wear and replacement if necessary, adjustment of engine controls and scheduled replacement of items such as oil filters and camshaft drive belts. All of these tasks are performed to a schedule determined by the rate at which the components wear or degrade in use. Repairs are performed either immediately (for example, engine fails to start or brake warning light on), or at some later time (for example, radio fails, fuel gauge fails).

Items that wear or degrade in use will fail at some time. If the time to failure is beyond the normal life of the product there is no problem: for example a bearing might have an expected life of 50 years in a system designed to work for a maximum of 20 years. If failure does not have serious effects, the part can be replaced or repaired when it fails: for example a light bulb in a display panel.

The decision on the best time to replace components, whether based on their condition, on a regular basis or on failure, should be made dependent on the following factors:

- The extent to which the onset of failure is predictable from knowledge of the component and its application. At one extreme are lighting units in streets or offices. The wearout life distribution is well known, as is the pattern of use. Also, it is much more economical to replace all units at one time rather than each unit on failure. Therefore it is easy to determine an optimum replacement interval, which ensures that only a small number of lights will fail before replacement of the lot. By contrast, at home we replace light bulbs and tubes on failure, because there is no economic advantage in wholesale replacement on a regular basis. Also, because the pattern of use varies, and the units presently installed will have varying ages, so there is no one optimum replacement interval for the whole house. Other situations are less clear-cut than these two extremes. For example, cyclically stressed components that might suffer fatigue damage and hydraulic hoses on earthmoving equipment have wearout lives that are much more widely variable and therefore harder to predict.

- Visible or other evidence of degradation can be used to determine replacement. For example, we can inspect tyres and flexible drive belts or perform non-destructive tests to detect cracks in structures. The component can be kept in use until such inspection or test indicates the need to replace it.
- The effect of failure must be considered. A leaking hydraulic hose on a grader is not critical, but rupture of a jet engine turbine disc is. Therefore the hydraulic hose can be replaced on failure, on condition or on a scheduled basis with a time interval that is not crucial. The actual decision would depend upon costs of downtime and of replacement. The turbine disc, however, must be replaced before it fails, and therefore its pattern of use (number of take-offs, flying hours) must be carefully recorded to ensure that the prescribed life is not exceeded. The prescribed life will be very conservative, typically a factor of at most half the expected life to failure.

Optimum maintenance planning must be based on knowledge of the underlying physical, chemical or other processes that cause degradation. Therefore maintenance planning must be made part of design, as discussed in Chapter 7. The great majority of electronic components suffer no degradation during normal use and over normally expected equipment lives. Therefore there is no benefit in testing or replacing electronic units before failure. However, if accuracy is important and if the user cannot observe changes, as is the case with most test equipment, then regular calibration must be performed. On the other hand, machine bearings always wear, the rate depending on load and application, so checking and replacement should always be considered.

Reliability Centred Maintenance (RCM)

These considerations are all part of the maintenance concept known as Reliability Centred Maintenance (RCM). RCM is used by operators of complex equipment such as airliners, industrial plant and military systems, when it is important to strike the optimum balance between availability, cost and safety. The starting point is a careful listing of all components that might be subjected to wear or degradation in use, with details of the relevant processes, life expectancy, extent of uncertainty in the life estimate, criticality and cost of failure, warning signs, inspection methods and replacement costs. The optimum maintenance and replacement policy, including inspection and replacement frequency, is then determined for each item

The task frequencies must be grouped into convenient intervals, since obviously it would not be cost effective for a complex product or system to be subjected to a large number of preventive maintenance tasks all at different frequencies. For example, car maintenance intervals are typically at 10 000 km: all high-frequency tasks (e.g. engine oil change, wheel alignment check) are performed every 10 000 km, lower frequency tasks (e.g. oil filter change, spark plug change) are performed every 20 000 km, and other tasks (e.g. brake seal change) every 40 000 km. Such grouping of tasks as multiples of the most frequent intervals minimises inconvenience and non-availability, and is therefore the least expensive. The same principle is therefore appropriate for other systems, such as process equipment, aircraft and railway systems. Suitable tolerances on the maintenance intervals are included, typically 10%, to provide flexibility.

It usually is not practicable to develop the optimum maintenance plan at the first attempt for a new product or system. Depending upon the extent of uncertainty regarding particular degradation processes, conditioned also by requirements for availability and safety, a conservative policy should be used initially and the maintenance intervals should be extended as evidence builds up to justify extensions. It is necessary to maintain careful records of the condition of the units checked or replaced and of any failures that do occur, in order to justify maintenance time extensions. Regular maintenance reviews should be held at which the engineers most closely involved with the items being reviewed must participate. Such personal participation by the people most closely involved, including if appropriate engineers from the suppliers of the items concerned, is essential, since the data recorded seldom provide all the information necessary.

Instead of grouping preventive maintenance tasks, it can be more appropriate in some cases to make use of other opportunities to perform them. Removing equipment from service just to perform preventive maintenance can be expensive, particularly if the maintenance is due at a time when the equipment is needed in service. If the pattern of use of the system is such that preventive maintenance tasks can be performed during periods when the system is not needed or is out of use for other reasons, including repair of failures, removal purely to perform large blocks of preventive maintenance tasks can be eliminated or minimized, thus improving availability. This approach should be considered for products such as military systems, transport fleets, agricultural machines, industrial plant and any others for which availability is important but the equipment is not required to be continually in service. The approach requires close control and recording to ensure that the best use is made of maintenance opportunities, and that all tasks are performed within the scheduled intervals.

Total Productive Maintenance (TPM)

Total Productive Maintenance (TPM) is an approach developed in Japan for optimizing the maintenance of production machinery in relation to its productivity. It integrates aspects of RCM and variation reduction techniques, and is applied by the people operating the machines using the same team-based approach as quality circles (described in the next chapter).

Managing Maintenance

Maintenance of engineering products can be as influential a factor as design and production in relation to operating costs, reliability and safety. Therefore maintenance must be managed with the same care as the upstream activities. People performing maintenance must be adequately trained and organized, and they must be motivated to play their part in minimizing costs and downtime. They should be involved in the design decisions that affect maintenance and in maintenance planning. Their experience should also be used in generating design improvements and in refining the maintenance tasks.

Maintenance is often performed by the user, for example by airlines, railways and factory and military equipment users, as well as by individual purchasers of products such as bicycles and power tools. When the user is an organization, maintenance of its engineering assets can be managed as described above, and there should then be a close and continuing relationship between the supplier and the user. Of course this is not practicable with large numbers of individual users, and then maintenance tasks must be made simple and the user handbooks must be written for the level of knowledge and skills of these users.

For many modern products, such as cars, office equipment and electronic test equipment, the supplier performs all or most of the maintenance. This is necessary when the product is too complex to be maintained by the user, apart possibly from simple inspection and replenishment tasks. For example, cars have become progressively more difficult for the average motorist to maintain, since specialist tools and diagnostic equipment are required for many tasks and owners do not have the knowledge to undertake them. For such products the speed and effectiveness of maintenance, both for repairs and for preventive tasks, is an important aspect of product reputation and success, and can be the deciding factor for many customers. For products in wide use, possibly internationally, such service must be based on a comprehensive and responsive data system which maintains full records of all items, including modifications, repairs and major preventive maintenance such as replacement of components subject to wear. Maintenance engineers must be trained to provide a service that will generate and retain customer preference for the product. The maintenance engineers are then much more than just repairers and adjusters: they become in effect part of the sales team, particularly when they come into frequent contact with users. Therefore they should be trained in this role and, whenever practicable, they should be given the opportunity to work directly with sales people.

Preventive and corrective maintenance tasks should be performed by the same people. Some organizations relegate preventive maintenance and the more simple repair tasks to lower grade people, reserving more complex repair and replacement tasks for more skilled people, possibly in a different part of the organization. However, separation of maintenance tasks in this way can lead to inadequate communication and lost opportunities for overall improvements in availability, reliability and costs. The total maintenance task should be organized as a team effort, with all levels of preventive maintenance and repair handled by the same people. In some cases it will not be practicable to apply this principle totally. For example, aircraft engine major repairs and overhauls cannot be performed on the aircraft so specialist facilities must be set up for this work, and electronic equipment usually requires special facilities for repair and test. The personal skills and experience needed for different technologies and levels of maintenance might also dictate that the same people cannot perform all tasks, but they should, as far as practicable, be organized and led as a single team, using common data, methods of reporting and training.

Technology and Maintenance

Maintenance requirements and methods are strongly influenced by developments in technology. Most recent developments reduce the need for maintenance but at the same time increase the complexity of the tasks that are necessary.

Mechanical engineering developments that influence maintenance include bearings that are sealed for life or that use self-lubricating coatings, therefore not requiring periodic lubrication, and the combination of improved materials and more accurate machining, resulting in less wear and thus reduced need to replace components such as piston rings and gears. In electronics the most important developments have been in the complexity of circuits, requiring the use of expensive automatic equipment for assembly, test and diagnosis, thus forcing centralisation of this work

Technology also provides opportunities for improved data handling and analysis. The ability to maintain comprehensive data on all items in use and of all maintenance work can greatly assist the task of optimizing maintenance. Engineers in the field can have instant information and on-line assistance, for example fault finding advice and pictorial information, and central planners have access to data on all tasks, spares consumption, reliability problems and trends and other management information. In some cases the data on utilization and operating conditions can be transmitted automatically from sensors and data processing built into the equipment, directly to the central computer via internet, telephone or radio links. Automatic links such as these can provide several benefits, including early warning of problems, full information on utilization and operating conditions and more accurate and timely data.

LOGISTICS

The support of engineering systems requites the provision of facilities for maintenance, spares, documentation, training, and possibly other services such as transport. The nature and extent of this support, for which the military term "logistics" is increasingly applied, obviously depends upon the type of system and its use. The logistics system for a personal computer would require only the provision of low-value spares and a repair service to support retailers. However, the logistics system for a fleet of aircraft or trains requires careful optimization of expensive spares, multi-location repair and overhaul facilities and expensive support equipment such as engine and electronic system test facilities. The logistics system for such products represents a significant proportion of the total cost, and its effectiveness greatly influences the overall economics of the system being supported.

Logistics support planning is most highly developed in military applications. US Military Standard 1388 describes in detail how logistics support analysis (LSA) tasks must be performed, including the provision of forecasts of utilization, reliability, repair and overhaul facilities and durations, costs, weights, skill requirements and other information. From these inputs the optimum quantities of spares and other facilities are calculated. Other organizations apply similar or simpler methods.

The methods by which resources can be optimized are well developed. They range from simple calculations based on utilization and expected consumption, for example of oil filters that must be replaced at scheduled intervals, to much more complex analysis for an item such as an engine which will require different levels of repair and overhaul, and whose reliability cannot be accurately predicted. The optimum requirements for complex, expensive items can be determined by using probabilistic optimization methods and with simulations based on Monte Carlo techniques, both run using appropriate software. These methods allow optimization to be performed over a range of options, such as overall cost constraints, allowable stock-out risks and variations of levels of repair. They also make it easy to perform sensitivity analyses, to determine for example the effects of changes in parameters such as reliability or repair time. Reference [18] describes these methods.

For fairly complex systems, the assumptions to which the outputs are most sensitive, and yet which are most difficult to predict, relate to reliability. We discussed reliability prediction in the last chapter. It is particularly important that the best, most informed estimates are used as inputs to logistics studies. The ranges of uncertainty must be analyzed, as well as the likely effects of any factors that can affect reliability, such as modifications and changes in operating conditions.

Logistics analysis must be made no more complex than the

uncertainty of the assumptions will justify. For example, there is little value in performing a very complex study, involving the predicted reliability, repair times and other assumptions for a large range of components of a system if many of the key assumptions are subject to wide uncertainty. The methods described in US Military Standard 1388 tend to be unduly and unrealistically complex. However, it is equally important not to use naively simplistic methods that can generate misleading outputs. For example, if it is known that failures might be related to age rather than occurring at a constant average rate, then using a constant average assumption will generate misleading forecasts of repair arisings. When the logistics system logic includes complicating features such as different repair levels and locations, or a mixture of scheduled and unscheduled maintenance, then simple analysis can hide important effects.

As an example of oversimplified analysis consider the operator of a vehicle fleet which used an engine which could fail in use, necessitating removal for repair, but which also required removal for scheduled overhaul at fixed intervals. The data showed a clear tendency for this type of engine to be more likely to fail as it approached the overhaul life. Furthermore, if a failed engine had consumed more than half of its scheduled overhaul life, it was repaired and overhauled, to start a new life. Otherwise it was only repaired, to continue to its overhaul life when reinstalled. The operator simply combined the average time between failures and the overhaul time, and divided this figure into the utilization rate to determine the forecast rate of repair/overhaul arisings. In the event, very few engines were removed from service, for either repair or overhaul, during the first two years of service. This was followed by a period in which the removal rate increased rapidly, and greatly exceeded the forecast. Nothing had changed as far as the engine or its utilization was concerned, but a naive formula had resulted in a gross error in logistics planning.

The correct approach to this situation would have been to use a simulation to evaluate the dynamic effects of the logic and of the assumptions, and then to plan for the forecast pattern of repairs and overhauls. This approach would have enabled the effects of changed logic to be assessed. For example, would it have been more economical to overhaul the engines when more than half the overhaul life had been consumed to failure, say 70%? What would have been the effect of a particular modification to reduce the rate of age-related failures? Would it have been more economical to extend the overhaul life or to buy more spare engines? How would the probabilities of stock-out and of vehicle availability be affected? Such questions cannot be analyzed with simple formulae, and yet the implications on cost and availability can be very high.

Logistics Support: In-House Or Outsource?

For systems that require a significant amount of support to keep them in operation, such as military systems, aircraft, rail systems (rolling stock and infrastructure), other vehicles, telecommunications and broadcast networks, gas and electricity networks, etc., outsourcing this task is a possible option. Some businesses have taken this route, whilst others retain all maintenance and logistics work in-house. What are the factors that should be considered in order to judge which way to go?

- The first consideration is the extent to which support and operation are integrated activities. At one extreme, support and operation of naval aircraft on an aircraft carrier (maintenance, spares, weapons, mission planning) are inseparable. At the other, support for a fleet of buses or for hospital equipment could be planned and performed so that there is little or no impact on system use.
- The second is the extent to which support is a core technology of the operating organization. Again, front-line military equipment requires front-line support, by people with the appropriate training, both technical and military. On the other hand, the people who plan and operate bus routes do not need to know how to service them, and hospital staff cannot usually be expected to repair and calibrate modern medical electronic systems.
- The third is the economic case. If outsourcing seems to be cheaper in the long term it is obviously worth considering. Sometimes this is the only aspect that is considered and support is outsourced on this basis. Not surprisingly, some of these decisions have turned out to be very expensive.

Some recent examples of correct and incorrect support decisions can provide pointers for situations when the options might seem to be more balanced.

- In the UK, nearly all railway infrastructure maintenance was outsourced by Railtrack, the company set up on privatisation to operate and maintain the track, signaling and stations*. The companies that were awarded the contracts had little or no previous experience of the work and almost no competent staff. Railway engineering people at all levels from technicians to managers had to leave Railtrack employment and join the contractors. Within a few years a major accident led to the discovery that across the country tracks were in a dangerous condition due to inadequate maintenance. The entire UK network was drastically slowed for over a year while dangerous track was replaced. In the meantime Railtrack had lost its competence at managing the system, since most of its engineers had joined the contractors or retired. The eventual cost of the decision was out of all proportion to the amount that the decision-maker thought would be saved. Railtrack was bankrupted. The awfulness of the decision was learned the hard way, and the successor to Railtrack, Network Rail, is taking maintenance work back in-house.

- It has become fashionable for train operating companies in the UK (London Underground, the many train companies and others) to outsource all support for their train fleets. In some cases this extends to the entire provision of trains and maintenance depots, even to the extent of deciding how many trains would be required to operate the timetable. Guaranteed levels of train availability and reliability must be achieved. The train builders have had to become train maintainers, and there could be problems resulting from unexpected problems, conflicting priorities and divided loyalties, especially with new, complex train systems. It will be interesting to see how this approach works in the long term.

- Telecommunications and radio and TV broadcast companies mostly maintain their network equipment. Operating and maintenance management are closely integrated at the control centres and the companies understand the technology, their networks and their geography.

- Most airlines maintain their aircraft. They see operation and maintenance as closely integrated activities, they have the experience and they understand the safety regulations. The aircraft manufacturers stick to manufacture.

---

* The question of whether separating operation and maintenance of the infrastructure from that of the trains was a good idea, bearing in mind the significant interactions between them, especially on modern high-intensity, high-speed lines, is debatable.

CONCLUSIONS

The discussion in this chapter shows how important it is for the people who will be involved in selling, using, maintaining and supporting engineering products to be involved at the early stages of design and development and throughout the evolution of the project. However simple the product, the factors that will influence its marketability and its costs in use must be identified and optimized. For complex systems such as aircraft, factory automation and military systems the costs of maintenance and support over the system's life can exceed the initial development and purchasing costs, so there is great scope for reduction in overall costs by paying early attention to these aspects. For intermediate products and systems such as road vehicles, medical electronics and office equipment improved maintainability and optimized support can reduce life cycle costs. This is an important aspect of integrated engineering that is often overlooked or under-resourced.

One practical way to ensure that support aspects for complex systems are managed effectively is for the purchasing organization to assign engineers to work with the supplier. They should be located at the supplier's premises for the duration of the design and development phases. Their tasks should be to provide information to the designers on expected use and maintenance, to support and review the design and development processes and to develop logistics information and plans. For example, they should assist with preparation of maintenance handbooks, obtain spares lists, determine likely reliability, utilization and other factors and generate optimum spares quantities and maintenance plans. When the product is put into service these engineers should form the nucleus of the support teams. This ensures that all aspects of support for the product are planned and provided for by the time it enters service, and that the support people have the knowledge to support it. It also provides the supplier with the information he needs to produce a realistic, supportable product.

The extent of maintenance and logistics planning and of purchaser involvement obviously depends on the type of product and its market. Military systems developed for a single customer require very detailed planning, which explains why military procedures for this work are usually very comprehensive. When products are bought and sold on the basis of support systems that already exist, as is the case with most commercial and personal equipment such as trucks and cars, the support package is usually predetermined and there is little or no need for the customer to be involved in its planning. However, for a wide range of intermediate products, such as factory automation, transport systems and process plant, an integrated approach, with engineers from the purchaser's and supplier's sides working as a team, is necessary to ensure that introduction to service and long term use are smooth and economic and that logistics problems are anticipated and minimised.

12

ENGINEERING IN SOCIETY

The business of engineering exists within the framework of society at large, at all levels from local to international. Engineering contributes enormously to society: nearly every product and service used by modern society depends upon engineering. Even non-engineering products and services such as food and banking depend upon engineering: engineering products are needed for food production, processing, packaging and retailing, and banks rely upon computers and telecommunications. At the same time, engineering depends upon the supply of the right kind of people and on the attitudes of society. Despite the benefits that engineering has bestowed, these attitudes are not universally favourable. Many people perceive engineering as being not entirely honourable, or as a profession that lacks the esteem and rewards of others such as entertainment, medicine, law and finance. Partly this is due to the fact that engineering is so wide ranging, from making electric toasters to designing spacecraft. Engineering is often perceived as being associated with militarism and with damage to the natural environment. These attitudes become reinforced by people such as teachers, journalists and politicians who influence others, particularly young people, when their future directions are being determined.

Engineering is hard work, both to learn and to practise. Many other professions seem to be easier, and this is a further reason why people turn away. The effort involved in becoming qualified in engineering, and in engineering employment, must be rewarded by pride in achievement and confidence in the future.

The future of engineering therefore depends upon how it is managed within society. Engineering managers should attend not only to the internal affairs of their organizations, but must be conscious of the local, national and international social pressures that influence the effects of their decisions. They should in turn work to shape these influences so that they develop favourably towards engineering, and to society as a whole.

EDUCATION

Education is the lifeblood of engineering. Future engineers are formed by the early exposure of children to science and mathematics. Their directions are influenced by their aptitudes, so those with average or above-average abilities in these subjects are potential engineers.

Teachers of mathematics and science are mostly qualified in these disciplines, so they must be expected to teach them well and to generate interest among their students. They are not in general expected to be good teachers of other subjects in which they are not qualified, even if the subjects are in some way related. Very few teachers are engineers. School curricula have in the past not normally included engineering as a subject, so there is little tradition of engineering education or experience in primary and secondary education in modern society.

Since modern economies depend to such a large extent on the supply of engineers, it has been fashionable in many advanced countries to try to stimulate this by introducing engineering, or "technology", into schools. This has generally been a flawed approach, for three reasons. First, teachers have been expected to teach topics in which they are not qualified or experienced. Second, and probably more significant, is that since engineering is based on science and mathematics, these basics must be mastered before they can be applied to the learning of engineering. Of course there are overlaps between what can be called science and what can be called engineering, but the practical content of schoolwork should be planned to illustrate principles and demonstrate basic applications, and not to be premature exercises in engineering design. Third, all time spent on "technology" must be taken away from teaching of basic principles. In Britain, for example, all school children are required to study "technology"*, regardless of their aptitudes or inclinations and despite the fact that few teachers are competent to teach the subject. This combination of unqualified teachers and misguided teaching has had the opposite effect to that intended: not surprisingly, young people are dissuaded from taking up an engineering career, and for those that do their formal engineering training begins with less knowledge of the fundamentals than they would have had under the traditional

---

* The term has come to be associated almost entirely with "information technology" (IT). Children are taught how to use PCs, but not much else that is technological.

teaching regime.

Of course it is appropriate for schools to encourage an interest in engineering, as they should for all later courses of study. However, this effort should not be part of the mainstream curriculum. If engineering topics are taught they should be dealt with separately, for example as voluntary activities. The fact that modern economies need engineers is not a rational justification for teaching children engineering. We also need doctors, architects and hotel managers, but we do not teach these professions in schools.

School courses that are not aimed at preparation for higher education, but at less academically demanding but equally necessary vocational training and work, can usefully include more practical training, since the students will be joining the world of work on completion, and their school education is not required to equip them to the same extent for further learning. There is a need to recognize that not all young people are capable of or enthusiastic about further education, and that the needs of neither group are best served by forcing them to be educated together once these needs are identified.

The effect of social pressures and inappropriate school teaching, as well as of recent "progressive" trends in school education, has led to a decline in the entry of young people, particularly the most able, into engineering. Engineering colleges and universities have had to reduce their entrance standards. They have had to expend more initial effort and time on teaching the basics of science and mathematics to make up for the deficiencies in school education. This leaves less time for teaching what should properly be taught at this level, thus inevitably reducing the knowledge on graduation.[*]

There is a need for a spread of engineering graduate knowledge, from those who will work at the frontiers between science and engineering in research and in new applications to those who will be mainly involved in the practical business of design, development, test, production and maintenance. Engineering also covers a wide range of topics, specializations and depths of specialization. Higher education must therefore provide this spectrum of output, whilst ensuring that there is no general difference in the esteem and status of the different courses and of the institutions providing them. However, there has been a trend in many countries, particularly in the United Kingdom, to place greater emphasis on academic aspects. The policy-makers and teaching staff perceive practical courses, and the institutions that offer

---

[*] In the UK this trend has continued steadily downwards during the 10 years since these words were written. Stupidity at work?

them, as being somehow of lower status than those offering academic degrees and a research programme. These institutions then attract the more talented students, many of whom then find that their courses offer less in the way of practical engineering than they would have expected. Few university engineering courses include more than fleeting training or experience in manufacturing methods such as workshop skills. Practically none teach testing or maintenance.

The "customers" for engineering training institutions are not the students, but the organizations that will employ them. The qualified students are the products on offer. Modern industry does not want all of its brightest young engineers to be trained to be researchers. It needs practical intelligence in design, development, production and support as well.

The training for engineers who will work at the "sharp end" must include sufficient practical work, such as design, manufacture and test of products, the use of manufacturers' product data and other data sources, and use of modern methods and equipment such as CAE systems and test and measurement equipment. Such courses should include less theory than the more academic courses. They should be supported by periods in industry and with industrially sponsored projects. The teaching staff should be engineers, whose advancement depends upon the quality of their products as perceived by their customers, and not upon the number of research papers they publish (they would not normally publish any). There should be no bias against institutes that offer such courses, particularly in government funding, simply because they do not conduct research. This is essential if they are to attract a suitable proportion of the most able teachers and school leavers.

Engineering also needs people with lower academic attainments, at the non-graduate, technician level. These are the people who leave school without university entrance qualifications, but who are needed in supporting roles such as test, maintenance and production. The typical qualification route is via apprenticeships or vocational courses, leading to proficiency in skills such as welding, machine operation, electronic test and repair, or wider qualifications. Technicians are as important to success, in most engineering fields, as are more highly qualified engineers. They are needed for the full practical realization and operation of most modern engineering products. A team of brilliant engineers might create designs for a new product or system but without technician support the design will not become a product.

Again, governments and industry have important roles to play in the supply of suitably trained technicians. Only government can create the framework of training centres, properly funded and staffed. They must also encourage industrial participation, since the trainees will be mostly people already in employment. This encouragement must include incentives to industry to employ and train apprentices, with the training programmes closely integrated with those of the local training centres. An excellent example of such an approach is the system of technician education in Germany, where all school leavers who do not undertake higher education must attend local technical schools one day a week, for up to 2 years, to obtain further qualifications. Employers are required to release them for this training. This contrasts with the situation in Britain, where school children are given "work experience" of 2 weeks, even if they intend to go on to higher education, and where post-school vocational training has been subjected to a long series of "initiatives" that last a few years, provide spasmodic training to some, then are cancelled or changed. Unfortunately, in the UK the provision of apprentice-based training has almost disappeared, as the government pursues elitist and damaging plans to increase university attendance at the expense of more useful training. As a result, German industry has benefited from a stable supply of well-trained technicians, whilst Britain suffers a chronic and worsening skills shortage.

There is a large overlap between the work and skills of engineers and technicians. It is not helpful to create artificial boundaries between them. A person who has studied at technician level and then become more qualified as a result of practical experience and further training is just as much an engineer as an honours graduate with little practical experience, and often more so. Therefore the status and career prospects of engineers must be based on their competence and experience, not just on initial qualifications, and they should all be given the opportunities and encouragement to advance their knowledge and skills.

Stability is a critically important factor at all levels of education. Students, parents, teachers, and employers can be disorientated and disaffected if they are not familiar with the education system. Instability and rapid change make it difficult to guide young people and to compare standards. As a result standards fall, even when the changes are supposed to improve them, and students and employers lose faith in them. It is far better to encourage progressive improvements in existing systems than to impose dramatic changes, yet governments have imposed sweeping changes and continue to do so in countries in the West. The results have almost always been a general depression in standards, morale and respect for education, particularly in science and engineering.*

Teachers, at all levels, are the developers of the future talents needed by society. Sadly, in some of the most advanced Western countries, particularly in Britain and the United States, the status and rewards for teachers do not attract the best**. Teachers have a more fundamental impact on the future well being of advanced societies, and of their relative competitiveness, than any other profession. Doctors, accountants and lawyers, and even engineers, do not influence the long-term quality of society to anything like the extent that teachers do. Teachers should enjoy the same status and rewards as other professions, and this is the only way to guarantee that teaching will attract the necessary share of the most talented. This is particularly important for teachers of science, mathematics and engineering, for whom prospects in industry are usually better than for those with qualifications in subjects such as languages or history. The common practice of remunerating teachers at a standard rate, regardless of specialization and usually regardless of performance, is an appropriate ingredient in a policy of industrial decline, which no government intentionally supports but some nevertheless have managed to achieve.

It should be possible for professional people such as scientists and engineers to move between other employment and teaching in order to bring experience to the classroom and the lecture hall. This is particularly important in engineering training, because of the practical nature of the subject and the pace of change. Unfortunately this flexibility is reduced because few good engineers are attracted by the lower rewards for teachers, particularly in schools and non-degree-level training institutes. The formal pedagogical qualifications demanded for working as a teacher, particularly in schools, is also a barrier to easier movement of experienced engineers into teaching.

The proper role of government in engineering is to ensure the provision of educated, motivated people. Education, starting from primary school and continuing through various forms of adult education, particularly vocational training and university, cannot be managed by industry. Engineering managers can obtain capital, plant and materials, limited only by their finances, so long as they operate successfully in a commercial economy. However, they cannot ensure

---

* The 10 years since this was written show that this self-destruction continues, with lessons unlearnt. More stupidity at work?
** UK government interference in how and what schoolteachers teach, and the loss of classroom discipline, have made teaching an unenviable career choice.

the provision of the talents they need, except by competing for the supply available. If the supply is limited it imposes long term but intangible constraints on all aspects of performance.

The constraints are long term for the simple reason that it takes a generation to influence the quality of education. They are intangible because even if managers can find the numbers of people they need, if the quality of their education has not developed their knowledge, skills and attitudes to their full potentials, the shortfalls cannot be measured in terms of inventiveness, productivity, and effectiveness of management. Their employers can and should continue to develop them, since there is no end to an individual's potential to learn, but industry cannot make up for widespread, long-term inadequacy in a nation's educational provision. Individual companies can ensure that they select and develop the best, but the burden on industry as a whole can seriously restrict national competitiveness. The relative decline of the UK and US engineering economies over the last 30 years or so has been due largely to the reduction in educational standards, particularly in relation to science and engineering.

More so than in any other profession, engineering education is never complete. Every engineer requires continual updating as methods, materials, technologies and processes change, as systems become more complex, and as multi-disciplined approaches become increasingly necessary. Therefore engineering education should always emphasize the need for continuation and should stimulate further development. Employers should ensure that their people are given the opportunities and motivation to continue to learn, for example by linking advancement to training. Many universities and other institutions provide excellent short courses as well as post-graduate degree courses. Engineering managers should be aware of which courses are most appropriate to their work and should make use of them and help to develop them.

Of all the issues in engineering management, education is the one with the longest time-scale. It also has the widest impact, since it affects the whole supplier chain. Employers have a duty to promote continuing education, but they should not be expected to teach people what they should have learned at school or college, and they should be able to expect that the people they employ will be numerate, literate and motivated to learn more. Companies operating where engineering education is excellent will enjoy the effects for a generation. Those working in countries in which standards of education have been allowed to fall will carry an extra burden until long after improvements are made.

PROFESSIONAL INSTITUTIONS FOR ENGINEERS

As with other professions, engineers are served by institutions that foster their members' interests by providing journals, training, lectures, conferences and other services. The institutions also provide recognition and status. To varying degrees they control professional standards and conditions of employment, though generally engineering institutions do not have as much influence over the employment of engineers as, for example, do the professional institutions serving doctors, accountants and architects. The professional institutions provide valuable focal points for industry and academia, and they influence and sometimes approve the curricula of engineering courses.

Engineers should be strongly encouraged to join their professional institutions and to participate in their work. Such involvement is an important aspect of continuation training and it demonstrates interest and awareness. It is right for employers to acknowledge and reward membership and they should encourage these activities. Engineering managers and specialists should contribute to, encourage and learn from their institutions.

Like other collections of people, professional institutions can become self-serving and can lose sight of their prime function of supporting and uniting the profession and its members. This tendency can be observed in Britain, where the engineering institutions impose entry requirements that greatly inhibit membership by non-graduates. They also create separate institutions for "technician engineers", thus institutionalizing the idea that engineers who do not have university degrees must be represented by inferior bodies, and should be kept separate from their colleagues with more academic training. Large numbers of separate institutions also serve different specializations, and there is considerable overlap between these and between branches of the main institutions. Complex arguments run regarding amalgamations, relative status (some can award "chartered" status to members, others not), and higher levels of association*. In Britain there is also the quaint but irrelevant practice of obtaining a Royal Warrant and coat of arms, and other traditional forms of recognition such as the formation of livery companies in the City of London and the Fellowship of Engineering, all of which are supposed to further the

---

* Some organizations will not promote engineers to management positions unless they have "chartered" status, regardless of their other, possibly more relevant experience and qualifications. Stupid?

interests of engineers and engineering. All of this factionalism and instability confuses public perceptions of engineering and reduces its esteem.

The interests of engineers and engineering are best served by the excellence of the profession, not by artificial restraints on institution membership, factionalism and the creation of bodies that many practising engineers cannot join or influence. The engineering institutions in the United States, Japan and continental Europe give stronger emphasis to the needs of their members, in contrast with the more nebulous aspirations of status and influence. The relative strength of engineering in Britain and in these other countries does not indicate that the British approach is best, but rather that a return to basic principles might be more appropriate.

MANAGEMENT TRAINING FOR ENGINEERS

Formal management training for engineers is often provided as part of undergraduate training, but this tends to be severely limited by the demands of the rest of the curriculum. This does not matter, since few engineers will be given management responsibilities soon after graduation. Therefore most management training is post-graduate.

It is common for management training needs to be seen to be common across disciplines, so that most management courses are general and not specific to retailers, insurers, accountants, engineers or other professions. Consequently most management courses teach the same topics, and this tendency increases with the level of the courses.

Of course there are common features in the management of different kinds of enterprise. They all depend upon people, they all have financial responsibilities, and they all operate within social, legal and other constraints. Therefore managers at the higher levels of business must be made aware of these aspects. There is also value in the people engaged in different disciplines learning together. The standard training approach to satisfy these needs is the Master of Business Administration (MBA), a qualification much sought after by ambitious managers. MBA courses are offered by a large number of universities and business schools. Many companies sponsor the courses and consider the qualifications necessary for higher management positions.

In general, MBA courses teach business administration. Whilst administration (organization, planning, finance, reporting, law, marketing, etc.) is a necessary aspect of good management, it is peripheral. Good administration can compensate for poor management in a static organization, but it cannot create teams and drive progress and improvement. This requires the more fundamental talent of management: leadership. Leaders are not created on one-year MBA courses, even if leadership is included as a topic in lectures and discussions.

MBA courses have been criticised for being too concerned with the administrative details of business, for having grown too rapidly and therefore for diluting the quality of the average course, and for not producing managers who perform better. MBA teaching is, to varying degrees, still under the influence of scientific management. Some MBA course directors are responding to these criticisms by reviewing the structure and content of their courses. However, the fundamental criticism stands: an MBA qualification does not make a manager. Administration can be bought. Management, as described in earlier chapters, must be nurtured and grown. It is notable that in Japan, where Drucker's teaching has been most widely accepted, there are no business schools and the MBA is not considered to be the route to the top.*

Hardly any MBA courses teach engineering management.** As discussed earlier, the management of engineering is, in very important ways, different and more difficult than managing other enterprises. There is greater need to create multi-disciplinary teams of skilled people, projects are much more dynamic, and the knowledge base grows rapidly. Very few MBA courses teach, for example, how to form engineering teams, how to select core technologies, managing testing, the modern approach to engineering production or the principles of variation and quality. Whilst an MBA qualification can be useful for an engineering manager, it should not be considered sufficient.

In addition to the full-time postgraduate MBA courses, a wide range of shorter courses are also available on particular management topics. These are offered by business schools and by specialist consultants and teachers. Not surprisingly these vary enormously in value and in relevance to particular management situations. Engineering managers should be aware of which courses are relevant and useful to them and to their teams.

---

* Written in 1993. The first Japanese business schools opened in 2002, and some of them are now ranked among the most successful in the world.

** Some universities now offer courses in Engineering Business Administration (EBA).

Effective and inexpensive training can be provided simply by reading the best books and journals on management. The books listed in the References, and many others, are all important reading for engineering managers.*** They present directly the teaching of the most influential pioneers of modern management, rather than the possibly diluted efforts of lecturers and consultants. More attention should be given to learning via reading than is usually practised in modern industry. Books and journals are much less expensive than training courses, and managers should give time to encouraging reading and to discussing the topics studied.

"GREEN" ENGINEERING

Protecting and improving the natural environment is a concern of many engineers and engineering companies. This concern is now also expressed in politics and in legislation. In many countries environmental pressure groups such as Greenpeace make life interesting for engineers involved with products and processes that are noisy, noxious, non-biodegradable, or nuclear. There is no doubt that the pressure groups have benefited society, but, like most pioneering movements, their targets and methods sometimes appear irrational to engineers.

The most notable achievements of the environmental movement as far as engineering is concerned, and which have involved enormous engineering effort and cost, have been the noise reduction of aircraft jet engines, the reduction of lead and other emissions from vehicle engines, and the cleaning of rivers and emissions from coal- and oil-fired power stations.* They have secured many other less dramatic improvements on behalf of society, and they have forced the legislation that governs environmental issues. They have also had a negative impact in frightening society and politicians concerning the safety of nuclear power, although even here they have forced the nuclear power industry to banish complacency in their operations.

*** There are also bad books: some recent offerings by business "gurus" have presented ideas that contradict the fundamental truths of the new management, and some books on quality, reliability and testing should not have been published.
* In 2004 we can add CO_2 and ozone-depleting emissions that influence global warming, and lead-free solder. In other respects the "greens" have generated fears that are less well grounded, such as about the effects of electromagnetic radiation from power lines and telephones.

All engineering activities in design, manufacture, operation and support must now address the environmental impacts, real and imagined. Products are being designed to use fewer processes that can result in ozone depletion. Cars and copiers are advertised to be recyclable, and copiers must be able to use recycled paper. The engineers involved must understand the environmental issues from the scientific point of view and in relation to public and market perceptions, and also relevant legislation and regulations. Many companies now appoint a manager for environmental issues to provide advice and training and to ensure that products and processes comply with environmental needs and laws.

We are on the verge of an environmental explosion similar to the quality "systems" explosion described in Chapter 10. Standards are being written* and armies of consultants, inspectors and auditors are being formed. There has for some time been evidence of environmental protection overkill, which if not curtailed will lead to higher costs to industry, higher prices and reduced competitiveness with little or no further public benefit. Unfortunately, the public position taken by engineers, as perceived by society, has sometimes been reactionary, so it is difficult for them to argue for less regulation and more commonsense control. Engineers and the bodies that represent them must be sympathetic to public anxiety, and must help to educate legislators, the media and the public without appearing to be defending vested interests or covering up problems.

SAFETY

The safety of engineering products and processes is another area of major social concern. When engineering seems to fail in its duty to protect, as in major incidents such as system or structural failure leading to a plane or train crash, the Chernobyl accident, the Piper Alpha oil platform fire and the Space Shuttle disasters, public fears are aroused and pressure increases for engineers to be subject to tighter surveillance and control.

Public perceptions of risk are not always as rational as engineers would like. When the risks are difficult to understand or quantify the gap in understanding is increased. For example, public fears of nuclear radiation are based, to a large extent inevitably, on ignorance and on the invisibility of the radiation. These fears are reinforced by the fact that even the specialists are uncertain about the long-term effects of

---

* ISO14000 was published in 1996.

radiation exposure. To a lesser extent there is also fear of software, manifested for example by "experts" declaring publicly, after crashes of aircraft that use computerized "fly by wire" flight controls, that software is intrinsically unsafe because it cannot be proved mathematically or by test to be safe in all possible circumstances. Exactly the same can be said about pilots and aircraft wings, or about train drivers and railway signals, or about oil production platform maintainers and valve actuators. However, people believe that they understand the behaviour of other people, and of artefacts like wings, signals and valves. Most people do not understand software, they cannot see it or touch it, so fear is easily aroused. Engineers know that software is intrinsically safer than people or hardware performing the same function, since software cannot degrade or wear out and it is not subject to variation: every copy is identical. Software can contain errors, but properly managed design and test can ensure that it is correct in relation to safety-critical functions, if not in every possible but remote eventuality*.

The safety of engineering products and processes has always been a valid concern, heightened by news of accidents and public perceptions. It took on much increased significance with the movement in the United States during the 1960s led by Ralph Nader and culminating in the legislation on product liability (PL). Similar legislation has since been enacted in Europe. Basically, PL removes the onus on an injured party to prove that the supplier or manufacturer of a product that caused death or injury was negligent. Now the supplier of manufacturer must prove that he was not negligent, and that he had applied all available methods to prevent the accident. There is no time or damage limit in the U.S. law, so a plaintiff can sue without limit even for injury caused by misuse of a product designed before the legislation was enacted. In Europe the law is more sensible, providing for retrospection only for 10 years and awards for damages limited to the amount actually suffered, not punitive damages based upon the defendant's ability to pay. PL has greatly increased the financial risks inherent in product design that can conceivably cause death or injury. This applies to electrical equipment, vehicles, health monitoring or sustaining equipment and many other products. Damage awards, particularly in the United States, are sufficient to put companies out of business, and some have ceased trading because they perceive the risks as being too high. It is unfortunate that PL has, again particularly

---

* As pointed out earlier in Chapter 10, software has not been the cause of any recent disasters.

in the United States, become a monster that stifles innovation and business, whilst being a bonanza for lawyers, so the overall benefit to society is questionable.

Designers and producers of engineering products and systems must be alert to the potential hazards and they must take all necessary precautions against them. They must be aware of the statutory requirements on safety in every market in which the products or systems will be used. They must retain records of all analyses and tests performed to ensure safety, so that they can demonstrate that they have taken every reasonable precaution against hazards. These precautions must include labelling, instructions, quality control and maintenance. Contingency plans must be in place for product recall in case hazards are discovered later.

The safety responsibilities of engineers also extend to people at risk in places of work. Every engineering manager in research, development, production, maintenance or operation nowadays carries heavy statutory responsibility for safety at work. Engineering managers and their team members can be personally, and even criminally liable for acts of omission or commission that lead to injury or death.

The employers of engineers must ensure that safety liabilities are minimized by providing training and a system that works to eliminate all foreseeable risks from products, processes and operations. For most engineering companies it is now essential that a manager is appointed to coordinate safety issues.

BUSINESS TRENDS

One of the major features of modern business that influences the task of managers is the continuing pace of company acquisitions, mergers and disposals. These often occur across national borders, as business becomes more multinational. They nearly always lead to displacement of people and force changes in organizations and methods. They present opportunities for some, but uncertainty and loss for others. Since the main justification for the moves is usually financial, changes are often forced down by new managers in order to generate quick savings. However, the long term success of the changes can best be assured by careful assessment of the strengths and weaknesses of all parts of the new business (people and processes as well as the conventional "due diligence" appraisal of intellectual property, industrial property, machinery, stock, finances, etc., and by applying the new management.

Another important trend is the transfer of manufacturing work overseas in order to reduce costs. This is an understandable move, but it can be unsettling to the important interface between design and production. It is essential that all of the effects are considered, not merely the easily quantified cost difference due to lower labour and other costs. If the products concerned are complex or new, so that some continuing design effort might be required to deal with production or support problems, engineers must be available to deal with these and they must retain close communication with the production centre.

Much of modern business is driven by short term pressures, and this seems to be an increasing trend. Companies feel threatened by competitors, possible acquirers and shareholders if they do not generate profits and share price growth. Some of the reward contracts for CEOs and other board level people motivate them towards short term greed rather than the long term good of the business. As a consequence of these pressures, companies shed staff and cut investments in training and research. Of course it is necessary to seek to operate the business as economically as possible, but, as Drucker emphasized, the main duty of top management is to ensure survival of the business, and this means taking account of the long term. Short term economies, especially when driven as campaigns with defined targets ("reduce staff across the board by 5%"; "cut capital spending by 20%"; "cut indirect expenses (training, travel, etc.) by 50%", etc.), are often applied, but they can be very damaging and expensive in ways that accountants might not appreciate. The financial benefits soon appear on the balance sheet, but the damage to the business in the longer term is less apparent.

There is an unfortunately "macho" attitude to much of modern management, largely driven by these pressures and trends. This seems to have been fostered by books and articles on management by fashionable "gurus", presenting panaceas and "paradigms". The contributions from the management training schools have not resisted this trend. Maybe this is tolerable in enterprises that are not critically dependent on skill, training, teamwork and long-term effort and investment. However, "macho" management of engineering is counterproductive and damaging.

POLITICS AND PEACE

Like most other fields of enterprise, engineering is influenced by politics. The environmental and safety issues just described are partly political. Monetary exchange rates, import quotas, tariffs, standards and regulations, political pressures and the states of importing and exporting economies all influence engineering, particularly when products, components and materials are bought and sold across national or trading zone boundaries. Company strategy must take all of these factors into account, even though some, particularly economic factors, are difficult to forecast. This is a field in which many companies, particularly smaller firms moving into export or import for the first time, might obtain worthwhile advice from consultants or government or industry advisory agencies. Large companies, particularly those already operating multinationally, should normally possess the necessary knowledge and sources of information, but special situations might justify the use of external help.

The greatest political change affecting engineering in recent history has been the ending of the Cold War. Companies that have been able to rely on a steady stream of defence-related business have had to undertake agonizing reappraisals of their situations and plans. Some major companies have quit defence work altogether or have sold off the defence parts of the business. Others have consolidated, sometimes in partnership with others, to ensue that they secure a good share of the reduced total business. Others have diversified into non-defence engineering.

Defence contracting was becoming less lucrative in the major Western economies even before the end of the Cold War as a result of tighter, riskier contract conditions and reducing defence budgets. Few modern government-funded defence development and production contracts now make money, so much depends upon the potential for export. However, export markets for defence equipment are much less certain than they used to be and competition is severe, as well as being highly political.

Defence engineering has always had the benefits of being at the leading edge of science and technology, and of being well funded. To some people it carries a stigma, as though the engineers concerned were in part responsible for the use to which weapons are put. Additionally, since many defence equipment companies have enjoyed symbiotic relationships with their few large customers, not all have adapted to the modern concepts of management and they have reputations for bureaucracy and inefficiency*. Nowadays defence

---

* In the UK the "smart procurement" initiative was introduced to improve efficiency and reduce costs. The effects seem to have been the opposite, according to a recent parliamentary report.

engineering carries the further problems of shrinkage and uncertainty about how far the process will continue.***

In this situation engineering managers must decide what is to be their vision of the future. Any of the three options mentioned above might be appropriate for a particular company. Those that opt out of defence work need think no more of it.

Those that decide to stay in and consolidate must plan to survive in the new situation. There will be fewer major contracts, spread over longer time-scales, and less production. The problems of retaining skills will therefore increase. The risks of cost overruns are high on most defence development projects because of the combination of high engineering risk and the need to bid low in order to secure the contract. Companies have found themselves in the Catch-22 situation of having to bid to stay in business, but knowing that a winning bid would be unprofitable. In this difficult situation there are two linked survival strategies.

The first is to adopt the modern management principles in all of their operations. For those companies that have not already done so, very large increases in engineering productivity, through all stages of a project, can be achieved. If a contract is won using cost estimates from previous projects, this improvement can greatly improve the financial performance by reducing time, effort and waste. Also, bids can be much more competitive if based on the expected performance under the new philosophy rather than on past performance.

The second approach is to maximize the potential for follow-on business, particularly production, in the home and overseas markets. To achieve this it is essential to ensure that the initial specification does not jeopardize future sales by including features that greatly increase risks and costs without conferring significant benefits. Alternatively, if such features are demanded by the customer, every possibility of developing lower cost options without these features should be pursued in parallel. Typical features that might be treated in this way are protection against nuclear, biological, and chemical (NBC) attack, electromagnetic pulse protection, requirements for specialist test equipment, and any performance requirement that seems to present high risks in regard to available technology and methods.

The Quality Function Deployment (QFD) method described in Chapter 7 can be used for this analysis. Whenever possible, the

---

** Do European countries really need submarines that carry nuclear missiles, or a new fighter aircraft?

purchasing organization should be invited to participate in this type of review and to agree with the decisions.

Defence equipment companies should also look for opportunities other than in new product development and production. Since defence equipment is expensive and remains in service for long periods, there are many opportunities for updating and improvement by applying new technology. For example, electronic units manufactured in the 1980s and 1990s can be replaced with smaller, lighter, cheaper, more reliable units using modern processing and packaging. Such improvements involve very little engineering risk and so can be developed and offered as direct replacements for the older units or as modification packages. Such possibilities are less obvious in non-electronic equipment, but the potential should be investigated for any older product that is expected to remain in use in sufficient quantities to justify updating or improving.

Companies that choose to diversify out of defence engineering must adopt the new management, if they have not already done so. Diversification will almost certainly put them in competition with world-class companies, which will already have achieved high levels of engineering productivity in development and in production. Therefore diversification without change will lead to failure, and this has been the fate of several past attempts.

Diversification must be based on the company's core technologies, and there must be a clearly defined product or market area that can be provided or served. Product managers must be appointed and given the freedom and encouragement to build teams and win. Only those procedures and policies that help should be retained: old ways and customs that existed as part of the defence contracting situation and which are not helpful in the new, such as complex modification approval procedures and functional reporting lines, should be swept away.

A few defence equipment companies in the USA and in Europe are now leading the way in coming to terms with the new realities of the post Cold War military equipment market. There will be more changes and the situation is still uncertain, so companies that stay in the defence business face interesting times, and those that move out must fight for position in markets that present them with new challenges.

NATIONALITY, CULTURE, AND GOVERNMENT

It is often stated that the economic rise of Japan since the end of World War II was primarily due to the differences in their society and culture which enabled them to impose conditions of work which would have been unacceptable to people in the West. Two facts demonstrate the error of this simple viewpoint. The first is that Japan lagged behind the West until the second half of the twentieth century, missing out on the early industrial revolution. The second is that the philosophy of management that led to the predominance of the leading Japanese companies was taught almost entirely by three Americans, Drucker Deming and Joseph Juran. Whilst Western society created these men, Japanese society created a large number of managers who listened and acted.

The story is not, however, simply one of permanent differences in attitudes of the people concerned. Powerful dynamic influences were also involved. Japan had been crushed and humiliated and two entire cities had been obliterated by atomic bombs. General Douglas Macarthur was appointed by the Allies to govern the country. One of his early acts was to break up the large industrial companies, just as was done in Germany under the Allied occupation. When later Deming and others were invited to teach, their audiences were not made up of the senior men who had led these companies before and during the war, but of younger men, facing new problems of management and reconstruction, and eager to learn from the country that led the world in productivity and quality. There was therefore a historical moment of receptiveness to ideas on management that were fresh and rational, and which promised a way towards reconstruction, growth and harmony between people at work.

There are features of Japanese culture that do encourage acceptance of the ideas of the new management, but it took the particular circumstances of the immediate post-war period to catalyse these. The main cultural difference in this respect is the more deductive, consensus-seeking approach to problems typical of most oriental society.

The deductive mentality is manifested in the realisation that opinions and methods are not necessarily either right or wrong, but that shades of grey exist and that improvement is always possible. Improvements can be generated by discussion, in which harmony is sought and confrontation is considered to be unmannerly. As far as possible disagreements are dealt with quietly and separately from wider gatherings. When problems are dealt with in this way progress can seem to be slow and frustrating by Western standards. However, progress is more likely to be in one direction, fewer mistakes are made, and everyone involved is committed to the solutions agreed.

This approach to dealing with problems is coupled with greater respect for age, experience and authority than is normal, for example, in America. General Macarthur and the American teachers he introduced were the new leaders, and the new young Japanese managers did not hesitate to follow them.

Western culture is not generally as consensus-seeking as is Japanese. It is marked by the Socratic philosophy of disputation as the way to resolve problems and the inductively-based philosophy that ideas and methods are either right or wrong. This attitude is reflected in Western parliamentary and legal systems, and it permeates Western management thinking. It is probably the greatest single obstacle to widespread and continuous acceptance of the new management principles by Western managers.

American industrial culture inherited the European philosophy. However, the American population is largely descended from the people who rejected the social systems of Europe, and who pioneered the building of a new nation. This stamped on American society an attitude of individualism and fierce entrepreneurism, coupled with distrust of authority. Working relationships are less formal and are based more on perceptions between individuals rather than on the positions they represent. These attitudes have encouraged the rapid formation and growth of new companies, formed to exploit new ideas. People move freely between companies, less constrained by loyalties or authority. Because of this fluidity and diversity, some companies eagerly apply the new management concepts, while others adhere to the Taylorist "scientific" principles. Management progress tends to be fitful and not always forwards, in the land of Drucker. Instead of steady progress across a broad front, there is a tendency to introduce specific "campaigns" and "initiatives", which often succeed only partially or fail, because they are not supported by an overall philosophy and long-term commitment. Consultants thrive on such conditions, offering advice, solutions and training on all of the current nostrums, or "paradigms". The risk-taking, free market, inductive American approach has led to the formation of companies that supply innovative technology used by the rest of the world, such as microprocessors, software and a wealth of other components, materials and systems.

Apart from products for which large financial outlays are necessary, such as military systems, aircraft, aero engines and cars, much of the development of new products has been conducted by companies that started small and grew rapidly. Competition on price, service and features is severe, often leading to financial strain or even bankruptcy in companies that supply excellent products, and market leaders can quickly fall behind. The smaller companies conduct little research and long-term development, of products or of people, though they are often closely linked with university research. Business life is hectic and exciting, and it is difficult to make forecasts.

European management, particularly in engineering, has generally been slower than the more competitive American firms to accept the new ideas. Also, the attitudes and structures have been much less conducive to the rapid formation and growth of new enterprises and movement of people. Consequently engineering developments in Europe, whilst not being behind in inventiveness, have not been exploited as effectively as American industrial culture exploits invention, or as effectively as Japanese culture exploits production. There have been high-flying start-ups, like Nokia, but the new technology markets have been dominated by big traditional companies like Siemens, Phillips, Thomson and GEC (though GEC has recently tumbled from its eminence, due to bad business decisions in relation to its core technologies).

The European approach tends to be conservative: fewer risks are taken and methods change more slowly. This is more marked on the European continent than in the UK, as continental management has been somewhat less influenced by American practices. Because of the cultural heritage, European managers are more prone to accept "systems" approaches to work, such as written standards and procedures. It is considered that it must be better to work in ways laid down by standards and procedures than in ways that are less "disciplined". Herein lies a fundamental contradiction: inductive work cannot proceed effectively when constrained by arbitrary methods and systems. Standards and procedures are incompatible with inventiveness, and they also limit improvements through deduction. Yet the legacy of "scientific management" leads to distrust and rejection of the fact that both inductive and deductive thought are more productive and better directed when not constrained.

Differences in national attitudes were reflected in the different policies adopted by Germany and Japan in order to boost production during the growth period of the 1960s. Germany imported large numbers of "guest workers" from Turkey and southern Europe, originally on short-term contracts but many were given residential status. The German "*Gastarbeiter*" were treated as unskilled, and were given minimal training. They were less educated than the German workers and were not even required to be proficient in the German language, so limiting the training they could receive. They were therefore destined to remain at the level of unskilled manual workers, with little possibility of advancement or of contributing to improvement. They could be managed only by "scientific" methods, being told what to do and how to do it. Even if they could, they had little motivation to improve the job, since they knew that (initially) their positions were temporary.

By contrast, Japanese industry had already adopted the new management. They used the concepts to extend automation, but in a human context. They developed the capabilities and knowledge of their indigenous production people so that they could control and improve the processes.

We can see that the German approach was based upon a short-term viewpoint: how to increase production using existing ideas and methods. The solution worked well in the short term, but the long-term consequences have been less favourable. The Japanese took the long-term view, which required investment in equipment and in people and the development of the new approaches to managing design and production that we have discussed earlier. The German approach was largely inductive: it was "a good idea at the time". The Japanese approach was deductive: it was developed by degrees, and all of the necessary actions and the consequences were thought out and agreed.

The moral here is that people should not be used merely as low-cost replacements or extensions for machines. Despite its strengths in other respects, it will take the German industrial economy a long time to recover relative to its competitors from this strategic error. I stress that the error was not in importing immigrant workers. It was in the way that they were treated: as low-cost assemblers and machine minders, different in kind from the skilled German workers. Such treatment might be appropriate for fruit-pickers or construction labourers, where changes are slow and skill and teamwork are not as important. It is not at all appropriate to engineering production.

Governments can help or hinder industry in other ways, such as by legislation on working hours, conditions of work, company law and taxation. These affect all industry, not only or mainly engineering. As a general philosophy, governments should not undertake roles that are the proper concern of managers. Examples of these are the European emphasis on enforcement of conditions of work and protection that are extremely expensive to industry in comparison with equivalent costs in other countries, and the American legislation on "positive discrimination". The problem with such legislation is that, once enacted, it is very difficult to change. Decisions on such matters, if taken by managers, can be based on consultation with their people and can be much more easily adapted to local circumstances and to changing conditions. Governments have a proper role to play in persuasion and in setting limits that are generally acceptable to society, such as the need for the workplace to be healthy and safe. However, interference at the level of management detail takes away from managers responsibilities that they should bear for the welfare and productivity of their people. Industrial competitiveness can be reduced, particularly in recessionary periods, resulting in reduced tax income from industry and higher welfare costs.

THE CHANGING WORLD OF ENGINEERING

The disintegration of the communist empire in Eastern Europe has released new opportunities for industrial growth. In most of the countries concerned, particularly the liberated European nations and Russia, there is a strong tradition of education and an established industrial base. Despite the presently grossly uncompetitive state of their industries, they retain the foundations of infrastructure, national organization and educated and experienced people on which to build modern engineering-led societies. They now face the challenge of exploiting this potential and reaping the benefits, whilst maintaining stability in the transitional period. Pressures of nationalism and population movements will continue to complicate these developments. It will be essential for the governments of these countries to maintain and develop the education and training of the people who will be needed to work in the new industries. Many of these will be set up by Western companies and there will be competition for these locations, so the availability of skilled people will be a strong inducement. It will also be essential for local managers to learn and to apply the new principles of management.

China is also emerging as a powerful new member of the world engineering community. Now that the grip of central state control is being relaxed and entrepreneurism encouraged, there is no reason in principle why China should not rapidly develop as an industrial and trading nation. The world knows the excellent contribution that expatriate Chinese have made to science and engineering. Despite the fact that a generation has suffered the educational repression of the "cultural revolution", the liberation of these talents in that vast country will present great opportunities for growth. Already many western and Japanese companies are becoming established in China, but the Chinese penetration of world markets for engineering products has yet to begin.*

Those parts of the world that were colonised by European states, such as much of Africa, south and central America and the West Indies, and parts of south-east Asia including India, still bear today a burden imposed by the colonising countries in relation to their attempts to develop engineering economies. Despite the fact that the colonising powers instituted education at school, and sometimes at university level, they did not export technical education. In departing, they left the countries they had controlled largely bereft of the infrastructure needed for developing their own engineering industries. Their very slow or negative economic growth during the second half of the last century has been largely due to the fact that they have had to import nearly all of the engineering products they have needed, and they have had to fund these from exports of produce and raw materials whose prices are largely out of their control. The fact that the governments of many of these countries also dissipated their limited resources on weapons and war has of course been a further burden to them. Now that the Cold War is over, so that international cooperation is more feasible, the industrialised world should begin to export the skills and training that the colonial powers so sadly neglected.

The Arab countries also possess combined potential for engineering development which has so far been insufficiently exploited. Again this has been largely exacerbated by preoccupations with military matters, which made the Middle East the world's greatest bazaar for weapons sales. Despite enjoying a common language and religion the Arab states have found cooperation difficult, and few have had stable, liberal governments for long enough to enable the development of an engineering culture. Most of the oil-rich states have not seen the need to do so. As the world economy moves forward again, adjusting to the new realities and shaking off the legacy of the Cold War, the development of engineering industries in the countries which had been largely left behind must be one of the major concerns of the governments concerned, and of the governments, institutions and major engineering companies of the more advanced industrial nations.

IN CONCLUSION: IS SCIENTIFIC MANAGEMENT DEAD?

Despite the wisdom of Drucker's teaching and the dramatic positive effects of its application, "scientific" attitudes to management still

---

* China has since become a major out-source manufacturer of engineering products, particularly electronics.

persist widely in Western industrial society. This is reflected in much of the modern literature and teaching on management and in the emergence of bureaucratic procedures, regulations and standards. Many managers, it seems, are unaware of the new management or are inhibited from applying it by the pressures and constraints of their work situation.

Engineering is the epitome of modern civilization. Like science and most art, it is truly universal. Like art, engineering is creative, and it can even create beauty. More than science and art it influences and changes all lives. However, not all of the results of engineering have been beautiful or beneficial, and the people engaged in its many forms are as human as the rest. Also, engineering is so widespread a discipline, and so much of it is perceived to be routine and unglamourous, or hardly perceived at all by the wider public, that its profession is often undervalued. Very few engineers achieve national or international recognition outside their profession. Even fewer have attained the kinds of reputations enjoyed by great writers, artists or scientists*. To a large extent this reflects the fact that few engineers work alone. In a world-class engineering product such as a high-speed train or a mobile telephone, every electronic component, plastic moulding and fastener is the product of engineering teamwork, as is every subsystem and the complete train system or telephone network. No credits are published, so the engineers' names are not listed as are, for example, the directors, producers, makeup artists, gaffers, focus pullers, sound recordists and others who contribute to the making of a movie or a TV film.

Engineering is therefore largely anonymous and the satisfactions are more personal, to the individual and to the team. Though engineers' names do not appear in lights they have the satisfaction that, in well managed teams and organizations, they can all influence the products and services in ways that are more fundamental than are allowed to gaffers and focus pullers.

Engineering and engineering management are not governed by any particular codes of ethics, as for example the Hippocratic oath taken by the medical profession. All professions are governed by law and by normally accepted standards of ethics. In addition, many engineering societies and institutions issue codes of practice, but these are not enforceable or supported by disciplinary frameworks. The profession of engineering has served society well without such additional

---

* Or writers and "artists" who create pretentious nonsense, or entertainers, or the generality of the modern "celebrity" class.

regulation, and there is no reason why it should not continue to do so and to improve its contributions. Since engineering management is a continuous, long-term activity, high ethical standards are essential for successful leadership and competitive progress.

All work should provide satisfaction to those engaged in it and to those who will benefit from it. However, mere satisfaction is barely sufficient. Work and its results should provide satisfaction beyond expectation, and happiness and fulfilment to those most directly involved. As Deming himself has stated, his famous 8th. point for managers ("drive out fear"), needs now to be stated as "create joy in work" Striving for happiness in work is not a naive or merely altruistic objective. It is the logical culmination of the most influential teaching of management. It is also the common observation of people at work, including the managers who make it happen, that the correlation between happiness and performance is very intense, yet subtle and fragile. Happiness can generate quality, inventiveness, teamwork and continuous improvements that transcend "scientific" plans and expectations. Happiness at work must be based on shared objectives and challenges, not on self-satisfaction or fear. It must be tempered by efficiency and discipline, and reinforced by learning and by the freedom and duty to contribute to the common effort. It must be of a quality that inspires and encourages effort, but not adulation or blind subservience. Generating and strengthening happiness at work, of such quality, and maintaining it in the whole of the organization despite the problems and changes that human enterprises must encounter, is the most difficult but most satisfying and rewarding task of managers.

This is the fundamental challenge for modern managers. In free societies, from which the fear of tyranny and global war has hopefully been removed, human creativity and productivity can be developed to their fullest when the work of individuals and of teams creates and increases happiness.*

---

* Ten years later, this message still seems apt. History has not ended, and the world still faces real and imagined dangers. I would merely add the words of (?): "Creating happiness is the first role of statesmen". I would also quote the words of the American Declaration of Independence about peoples' "unalienable rights" to "life, liberty, and the pursuit of happiness." Creating happiness is the first role of managers, at all levels.

REFERENCES AND BIBLIOGRAPHY

[1] P.F. Drucker, *The Practice of Management*, Heinemann 1955.
[2] F.W. Taylor, *The Principles of Scientific Management*, Harper and Row 1911.
[3] A.H. Maslow, *Motivation and Personality*, Harper and Row 1954.
[4] D. Mcgregor, *The Human Side of Enterprise*, McGraw-Hill 1960.
[5] W.E. Deming, *Out of the Crisis*, MIT University Press 1986.
[6]P.M. Senge, *The Fifth Discipline: the Art and Practice of the Learning Organization*, Doubleday 1990.
[7] T.J. Peters and R.H. Waterman, *In Search of Excellence*, Harper and Row 1982.
[8] Rosabeth Kanter , *The Change Masters*, Simon & Schuster, 1985
[9]J.P. Womack, D.T. Jones, and D. Roos, *The Machine that Changed the World*, Macmillan 1990.
[10] G. Hamel and C.K. Pralahad, *Competing for the Future*, 1994.
[11] B. Thomas, *The Human Dimension of Quality*, McGraw-Hill 1995.
[12] K. Murata, *How to Make Japanese Management Methods Work in the West*, Gower 1991.
[13] D. Clausing, *Total Quality Development: a step-by-step guide to World Class Concurrent Engineering*, ASME Press 1994.
[14] Donald G. Reinertsen, *Managing the Design Factory*, The Free Press 1997.
[15] P.D.T. O'Connor, *Test Engineering*, John Wiley and Sons 2001.
http://www.pat-oconnor.co.uk/testengineering.htm
[16] Masaaki Imai, *Kaizen: The Key to Japan's Competitive Success*, McGraw-Hill/Irwin 1986. (and his 1997 book *Gemba Kaizen* ("*Kaizen in the workplace*")).
[17] W.A. Shewhart, *The Economic Control of Quality of Manufactured Product*, Van Nostrand 1931.
[18] D. Hutchins, *In Pursuit of Quality*, Pitman 1990.
[19] P.D.T. O'Connor, *Practical Reliability Engineering* (4th. edition), John Wiley and Sons 2002.
 http://www.pat-oconnor.co.uk/practicalreliability.htm

The following books are also interesting and instructive:
J.L. Adams, *Flying Buttresses, Entropy and O-rings*, Harvard University Press 1991.
S.L. Florman, *The Existential Pleasures of Engineering*, St. Martin's Press 1976.
B.R. Rich and L. Janos, *Skunk Works*, Warner Books 1994.

INDEX

D

E